Graphing Calculator Manual for the TI-83 Plus, TI-84 Plus, and TI-89

Patricia Humphrey
Georgia Southern University

to accompany

The Triola Statistics Series:

Elementary Statistics, Tenth Edition

Elementary Statistics Using Excel, Third Edition

Essentials of Statistics, Third Edition

Elementary Statistics Using the Graphing Calculator, Second Edition

Mario F. Triola
Dutchess Community College

Boston San Francisco New York
London Toronto Sydney Tokyo Singapore Madrid
Mexico City Munich Paris Cape Town Hong Kong Montreal

Reproduced by Pearson Addison-Wesley from electronic files supplied by the author.

Publishing as Pearson Addison-Wesley, 75 Arlington Street, Boston, MA 02116.

ISBN 0-321-36920-3

4 5 6 BB 09 08 07

Preface

Texas Instruments (TI) graphing calculators are wonderful tools to aid in the study and implementation of statistics. They have many built-in features especially for the practice of statistics, features which this companion will introduce and explain. TI-83/83+/84/84+ calculators work identically for most procedures (with some minor menu changes, depending on the operating system version for your particular calculator). The TI-89 series has a different language and menu system which may seem incompatible at first, but soon becomes transparent. Differences between the two series will be pointed out and separate subsections will handle the different models, at least until a point in the course where a student should be familiar with his or her own model. This companion is designed to be used side-by-side with the text *Elementary Statistics 10th Edition* by Mario F. Triola. As an example, Chapter 2 of the text introduces graphs of data. It describes various types of graphs as well as summaries of data and uses an example based on the ages of winning Best Actresses. This companion uses the <u>same</u> examples as the text to illustrate how the TI calculators can be used to make the various types of graphs.

This companion follows the text of *Elementary Statistics* chapter by chapter offering TI techniques for most topics. The examples used are directly from the text. These examples are often paraphrased for the sake of brevity, but should be recognizable enough to the reader. The reader is urged to consult the text for important details such as definitions, requirements for running a procedure, and detailed conclusions in context.

Chapters 1 and 2 are very important for your understanding of all the other chapters in this companion. Chapter 1 introduces you to the TI keyboard and to some skills which you will need often, such as saving and deleting data. Chapter 3 introduces you to the TI calculator's capabilities in descriptive statistics.

There are undoubtedly a variety of ways to perform the operations described in this companion. In most cases, I have chosen to present only one way (hopefully, the easiest) of doing everything.

I have chosen, in most instances, to show exact key strokes for all instructions along with a narrative. I strongly urge you to look at all times at what you are pressing and the result. If you hurriedly press a sequence of five or six buttons without paying attention, you may get the desired result, but it will be hard for you to duplicate on your own because you will not see the logic behind the key choices. It is valuable for you to know where things are located on the TI menus.

I wish to thank Mario F. Triola, without whom this companion would not be possible. I also wish to thank the many students who have taken my statistics courses over the past nine years at Georgia Southern University. Thanks also to Deirdre Lynch and Joe Vetere of Addison Wesley for their patience and assistance, and to Sharon Taylor of Georgia Southern University for her careful proofreading.

Contents

1 Introduction to Statistics (and TI Graphing Calculators)

In this chapter we introduce our calculator companion to Triola's *Elementary Statistics* (10th ed.) by giving an overview of Texas Instruments' graphing calculators: the TI-83, -83+, -84+, and -89. Read this chapter carefully in order to familiarize yourself with the keys and menus most utilized in this manual

You will also learn how to set the correct MODE on the calculators to ensure that you will obtain the same results as this companion does. You will learn other useful skills such as adjusting the screen contrast and checking the battery strength

Aside from the above technical skills, you will learn some basic skills that are particularly useful throughout your study of Triola's *Elementary Statistics*. Throughout this companion, we will present the uses of these calculators by illustrating their use on actual textbook examples. The first will be an exercise from the text which requires the selection of a random sample. There are not many calculator exercises in Chapter 1 of your textbook because it is an overview and introduction chapter. We will take the opportunity in Chapter 1 of this companion to introduce you to skills which you will find necessary throughout the other chapters. These skills include Home screen calculations and saving and editing lists of data in the STAT(istics) editor.

KEY DIFFERENCES BETWEEN THE 83/84 SERIES AND THE 89 SERIES

All calculators in the TI-83/84/89 series have built-in statistical capabilities. Although a few statistical functions are "native" on the TI-89, most of the topics covered in a normal Statistics course require downloading the Statistics with List Editor application which is free. Download requires the TI-Connect cable. See the web page http://education.ti.com/us/product/tech/89/apps/appslist.html for more information. This manual assumes the statistics application has been loaded on the calculator. If you have the newer TI-89 Titanium edition, the statistics application comes pre-loaded, and the TI-Connect cable is included with the calculator.

The TI-83 and -84 series calculators are essentially keystroke-for-keystroke compatible; however, the 84 does have some additional capabilities (some additional statistical distributions and tests, for example) with the latest version of the operating system, version 2.30 which is also available for download at education.ti.com. The regular TI-83 (not the plus version) does not have the ability to use APPS (applications) which are in some cases extensive programs. If you have one of these regular calculators, you will not be able to use the APPS included on the CD-Rom accompanying the text to load data sets, but will have to key them in yourself. (If you have the cable and TI-Connect software, these can be loaded from the .txt files included on the CD – see the Appendix for details). Regular TI-83 users will, however be able to use the programs on the CD for such applications as analysis of variance and multiple regression.

There are some key differences in calculator operation and menu systems which will, in some cases necessitate separate discussions of procedures for the TI-83/84 and TI-89 calculators. Some of these become apparent in the next section. Not only are there differences between the three series, but there is a key difference in operation between the TI-89 and the TI-89 Titanium edition. On the Titanium, all "functions" on the calculator are essentially applications – when a regular TI-89 is turned on, the user is on the "home screen" similar to that for the TI-83 and -84. When the Titanium edition is first turned on, one must scroll using the arrow keys to locate the desired application – we'll say more about this later.

KEYBOARD AND NOTATION

All TI keyboards have 5 columns and 10 rows of keys. This may seem like a lot, but the best way to familiarize yourself with the keyboard is to actually work with the calculator and learn out of necessity. The keyboard layout is identical on the 83+ and 84+, and differs from the 83 by the substitution of the [APPS] key for the [MATRX] key. The layout of the 89 (and 89 Titanium) keyboard is similar, but some functions have been relocated. You will find the following keys among the most useful and thus they are found in prominent positions on the keyboard.

- The cursor control keys [◄], [►], [▲] and [▼] are located toward the upper right of your keyboard. These keys allow you to move the cursor on your screen in the direction which the arrow indicates.
- The [Y=] key is in the upper left of the 83 and 84 keyboards. It is utilized more in other types of mathematics courses (such as algebra) than in a statistics course; however you will use the *yellow* [2nd] function above it quite often. This is the STAT PLOT menu. We will discuss the yellow [2nd] functions shortly. On TI-89 calculators, the Y= application is accessed by pressing [♦][F1].
- The [ON] key is in the bottom left of the keyboard. Its function is self-explanatory. To turn the calculator off, press [2nd][ON].
- The [ENTER] key is in the bottom right of the keyboard. You will usually need to press this key in order to have the calculator actually do what you have instructed it to do with your preceding keystrokes.
- The [GRAPH] key is in the upper right of the keyboard. On the TI-89, GRAPH is [♦][F3].

As mentioned briefly above, most keys on the keyboard have more than one function. The primary function is marked on the key itself and the alternative functions are marked in color above the key. The actual color depends on the calculator model. Below you will be instructed on how to engage the functions which appear in color.

The [2nd] Key

The color of this key varies with calculator model. On 83's and 89's this is a yellow key near the top left. On 84's and the 89 Titanium, the key is blue but is also at the top left. If you wish to engage a function which appears in the corresponding color above a key, you must first press the [2nd] key. You will know the second key is engaged when the cursor on your screen changes to a blinking ■. As an example, on a TI-83 or -84 if you wish to call the STAT PLOTS menu which is in color above the [Y=] key, you will press [2nd] [Y=].

The [ALPHA] Key

You will also see characters appearing in a second color above keys which are mostly letters of the alphabet. This is because there are some situations in which you will wish to name variables or lists and in doing so you will need to type the letters or names. If you wish to type a letter on the screen you must first press the [ALPHA] key. The color of this key depends on the model of calculator: on TI-83's it is blue; on 84's, green; on the 89, purple; and on the 89 Titanium, white. On 83's and 84's it is directly under the [2nd] key; it is one place to the right of that on both 89's. You will know the [ALPHA] key has been engaged when the cursor on the screen turns into a blinking ■. After pressing the [ALPHA] key you should press the key above which your letter appears. As an example if you wish to type the letter E on an 83 or 84, press [ALPHA] [SIN] (because E is above [SIN]. To get the same letter E on an 89, press [ALPHA][÷].

Note: If you have a sequence of letters to type, you will want to press [2nd] [ALPHA]. This will engage the colored function above the [ALPHA] key which is the A-LOCK function. It locks the calculator into the Alpha mode, so that you can repeatedly press keys and get the alpha character for each. Otherwise, you would have to press [ALPHA] before each letter. Press [ALPHA] again to release the calculator from the A-LOCK mode.

Some General Keyboard Patterns and Important Keys

1. The top row on 83's and 84's is for plotting and graphing. On 89's these functions are accessed by preceding the desired function with [♦].
2. The second row down has the important QUIT function ([2nd] [MODE] on 83's and 84's, [2nd][ESC] on 89's). On 83's and 84's it also contains the keys useful for editing ([DEL], [2nd] [DEL] (INS), [◄], [►], [▲] and [▼]). INS and DEL on 89's are both combination commands: INS is [2nd][←] and DEL is [♦][←].
3. The [MATH] key in the first column on 83's and 84's leads to a set of menus of mathematical functions. Several other mathematical functions (like [x^2]) have keys in the first column. On a TI-89, [2nd][5] leads to the Math menu.
4. The keys for arithmetic operations are in the last column ([÷] [×] [−] [+]).
   Note: On all input screens, the [÷] shows as /, and the [×] shows as *. On both 89 models, when the command is transferred to the display area the * is replaced with a · and division looks like a fraction.
5. The [STAT] key, on 83's and 84's will be basic to this course. Submenus from this key allow editing of lists, computation of statistics, and calculations for confidence intervals and statistical tests. On 89's with the statistics application, one starts the application using the key sequences [♦][APPS] and selecting the Statistics application. On the 89 Titanium, quit the current application ([2nd][ESC]) and locate the Stats/ListEd application. On 83's and 84's the second function of the [STAT] key is LIST. This key and its submenus allow one to access named lists and perform list operations and mathematics.
6. The [VARS] key on 83's and 84's allows one to access named variables. On TI-89's this is [2nd][-] which is named [VAR-LINK], which is used for both lists and variables.
7. [2nd][VARS] calls the distributions (Distr) menu. This is used for many probability calculations. To get this menu on a TI-89, press [F5] from within the Stats/ListEd application.
8. The [,] key is located in the sixth row directly above the [7] key on 83's and 84's, while on 89's it is above the [9] key. It is used quite often for grouping and separating parameters of commands.
9. The [STO►] key is used for storing values. It is located near the bottom left of the keyboard directly above the [ON] key on all the calculators. It appears as a → on the display screen.
10. The [(-)] key on the bottom row (to the left of [ENTER]) is the key used to denote negative numbers. It differs from the subtraction key [−].

Note: The [(-)] shows as ⁻ on the screen, smaller and higher than the subtraction sign.

SETTING THE CORRECT MODE

If your answers do not show as many decimal places as the ones shown in this companion or if you have difficulty matching any other output, check your MODE settings. Below we instruct on setting the best MODE settings for our work. These are the ones we have used throughout this companion.

On an 83 or 84, Press the [MODE] key (second row, second column). You should see a screen like the one on the right. If your calculator has been used previously by you or someone else the highlighted choices may differ. If your screen does have different highlighted choices use the [▲] and [▼] keys to go to each row with a different choice and press [ENTER] when the blinking cursor is on the first choice in each row. This will highlight the first choice in each row. Continue until your screen looks exactly like the screen to the right. Press [2nd] [MODE] (QUIT) to return to the Home Screen.

On TI-89's, the default mode is to give "exact" answers. For statistical calculations, you will want to change the mode to give decimal approximations. To set this option, press [MODE]. Press [F2] to proceed to the second page of settings, then arrow to Exact/Approx and use the right and down arrows to change the setting to `3:Approximate`. Press [ENTER] to complete the set-up. The sequence of screens is shown below.

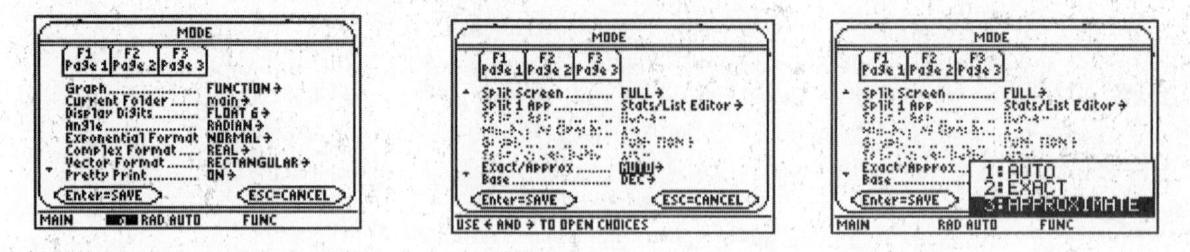

SCREEN CONTRAST ADJUSTMENT AND BATTERY CHECK

To adjust the screen contrast, follow these steps:

To increase the contrast, press and release the [2nd] key and hold down the [▲] key. You will see the contrast increasing. There will be a number in the upper-left corner of the screen which increases from 0 (lightest) to 9 (darkest).

To decrease the contrast, press and release the [2nd] key and hold down the [▼] key. You will see the contrast decreasing. The number in the upper-left corner of the screen will decrease as you hold. The lightest setting may appear as a blank screen. If this occurs, simply follow the instructions for increasing the contrast, and your display will reappear.

When the batteries are low, the display begins to dim (especially during calculations) and you must adjust to a higher contrast setting than you normally use. If you have to set the contrast setting to 9, you will soon need to replace the four AAA batteries. With newer versions of the operating system, your calculator will display a low-battery message to warn you when it is time. After you change batteries, you will need to readjust your contrast as explained above.

Note: It is important to turn off your calculator and change the batteries as soon as you see the "low battery" message in order to avoid loss of your data or corruption of calculator memory. Change batteries as quickly as possible. Failure to do so may result in the calculator resetting memory to factory defaults (losing any data or options which have been set).

A SPECIAL WORD ABOUT THE TI-89 TITANIUM

On the TI-89 Titanium, most all important functions which on other calculators are accessed by keystrokes, are applications (`Apps`). When the calculator is first turned on, you will be presented with a graphical menu of these applications, as at right. Paging through the screen to find the one you want can be tiresome and time consuming. There is a way to customize this screen so that you only see those applications you want to see.

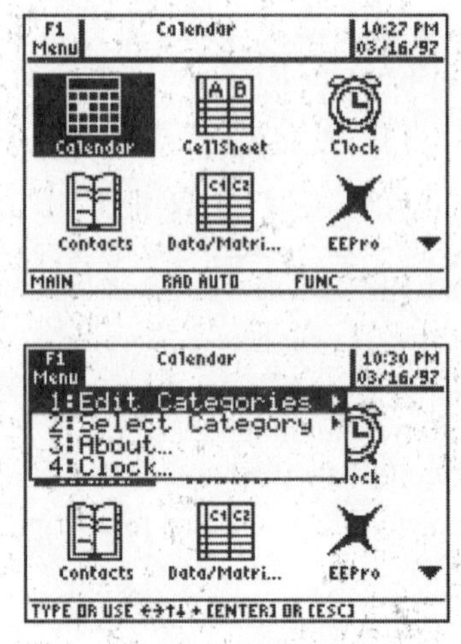

On the screen above, press [F1]. Press the right arrow key to expand menu selection `1:Edit Categories`. You will be presented with a list of possible catefories. Press [3] to select option `3:Math`.

On this screen, use the down arrow to page through the list of apps. When you find one you want to be displayed, press the right arrow key to place a checkmark in the box. The screen at right shows that the Data/Matrix Editor and the Home screen have been selected. For ths statistics course, you will want these apps, along with the Stats/List Editor and Y= apps. Press [ENTER] when you have finished making your selections.

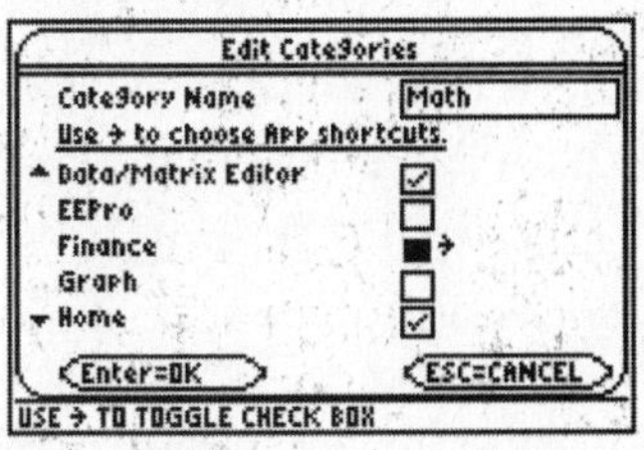

On this calculator, pressing [2nd][ESC] (Quit) will return you to the apps selection screen. There are two useful shortcuts between apps. The first is pressing the [HOME] key, which takes you directly to the Home screen. The other useful shortcut is pressing [2nd][APPS] which allows you to toggle between two apps.

RANDOM SAMPLES

All TI calculators have random number generators built into them. Such programs are dependent upon a value known as the "seed". Every calculation you make resets the seed, but we can set a particular value so that the sequence of random numbers will be the same. In practice, when one wants truly "random" numbers, one would eliminate this step. In this companion, we reset the seed each time we are generating a random sample, so that your output will match that in the companion. In normal practice, you do not need to reset the seed.

Random Sample and Simple Random Sample.

Picture a classroom with 60 students arranged in six rows of 10 students each. Use the calculator to simulate the results when a sample of 10 students is chosen as follows:
a) The professor will roll a six-sided die in order to choose a row. The students in that row are the sample.
b) The professor will choose a *simple random sample* of 10 students from the class.

TI-83/84 Procedure

Press [ON] to turn the calculator on. A cursor should be blinking on the Home screen. If not, press [2nd] [MODE] to Quit and return to the Home screen.

Press the [CLEAR] key if the cursor is not in the upper left corner.

We will first answer part a of the question by simulating the roll of a six-sided die. This means we need to generate a random integer between 1 and 6. Let us set the seed this time as follows:

a) On the Home screen type 123. Then press [STO▸] [MATH] [◄]. Watch what is happening – You went to the Math, PRB menu. Look at option 1. It is rand. You should see the screen at right.

b) Now press [1] and you will see that rand is pasted at the top of the Home screen as in the screen below.

c) Press [ENTER] and you will see the 123 as on the second line. This indicates the seed is now set.

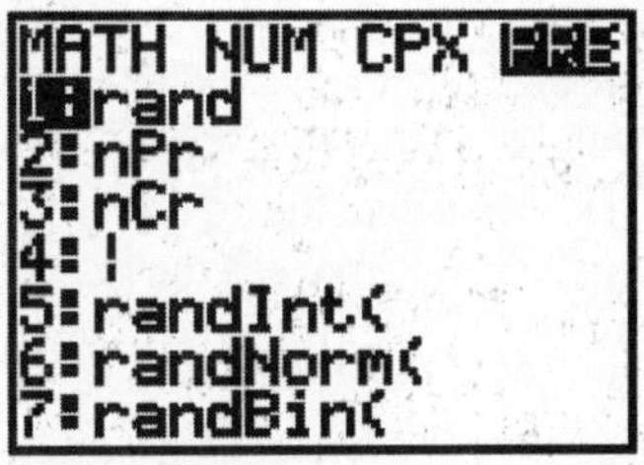

We will now simulate a single roll of a six-sided die by asking for a random integer between 1 and 6. Do so by pressing [MATH] [◄] [5] (thus choosing the option randInt(from the Math, PRB menu). Type 1,6 (remember the [,] key is above the [7]). Press [ENTER] to see the result. We see the result is a 5, so the students in row 5 would be chosen.

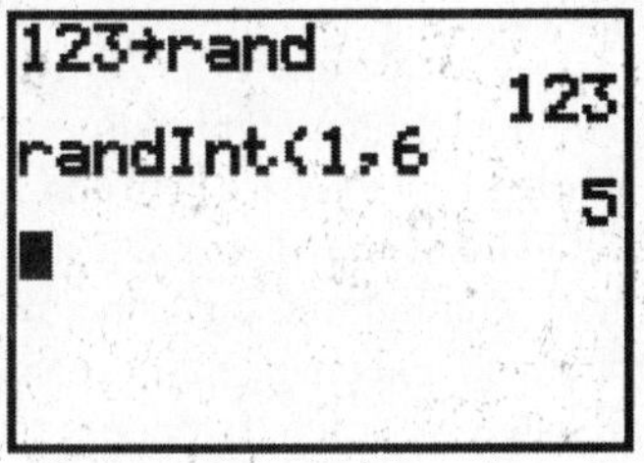

Now for part b of the question. We want a simple random sample of 10 of the 60 students. First number the students (1 to 60). Now you need a random sample of 10 integers between 1 and 60. This time we set the seed as 222. (This is only necessary if you wish to replicate our results.) After resetting the seed, press [MATH] [◄] [5]. Type 1,60,10. Press [ENTER]. You should see the screen at right. We cannot see all 10 numbers. You can use the [►] key to move across and see the ones which are not displayed. Note that the 10 we typed indicated we wanted a sample of size 10. In the above example, students 31, 1, 2, 11, 27, 52, 47, 4, 15, and 46 were chosen. It is possible that a student's number could have come up twice in the sample. If this occurs in a setting where this is not allowed (as in this one – a student should only be selected once), simply generate some more numbers until you have the desired sample size.

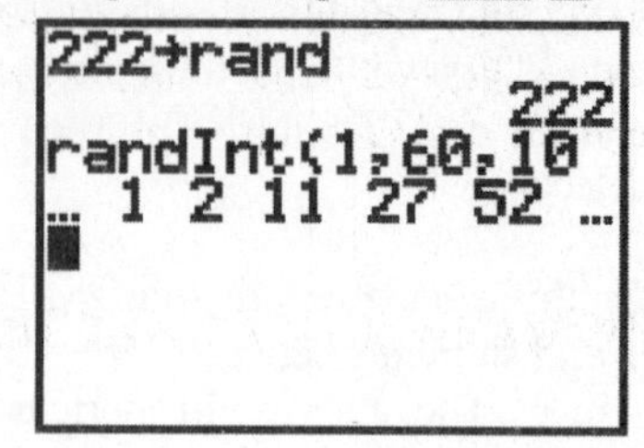

TI-89 **Procedure**

Load or select the `Stats/ListEd` application. Place the cursor at the top of an empty list. Press [F4] (`Calc`).

We will first answer part a of the question by simulating the roll of a six-sided die. This means we need to generate a random integer between 1 and 6. Let us set the seed as follows:

Arrow to `4:Probability` and press the right arrow to display the submenu. The menu option to set the seed is `A:RandSeed`. The easiest way to get there is to press the up arrow. Once the cursor highlights this option, press [ENTER] to select it.

Type the desired seed number. Here, I have used 123. Press [ENTER] twice to store the seed and return to the list editor.

Press [F4] again, and display the `Probability` menu. This time we want to select menu option `5:randInt(`. Now type 1,6) (remember the [,] key is above the [9] and we must close the parentheses). Press [ENTER] to see the result. We see the result is a 5, so the students in row 5 would be chosen.

Next we tackle question b). We want a simple random sample of 10 of the 60 students. First number the students (1 to 60). Now you need a random sample of 10 integers between 1 and 60. As above, we set the seed to be 222. (This is only necessary if you wish to replicate our results.) With the cursor highlighting the name of the list, we now again find the `RandInt` command and change the parameters to 1,60,10) and press [ENTER]. This indicated we want 10 random numbers between 1 and 60.

Pressing [ENTER] populates the list. In the above example, students 31, 1, 2, 11, 27, 52, 47, 4, 15, and 46 were chosen. To see the entire list, scroll through it to the end. It is possible that a student's number could have come up twice in the sample. If this occurs in a setting where this is not allowed (as in this one), simply generate some more numbers until you have the desired sample size.

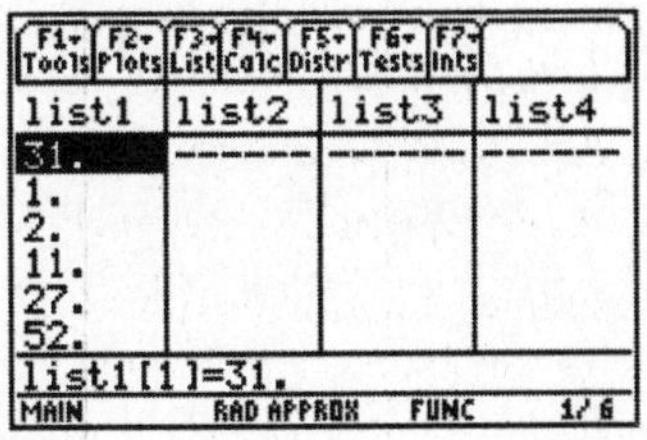

HOME SCREEN CALCULATIONS

Cumulative Review Exercises: Calculator Warm-ups

We will use exercises 2 and 5 of the cumulative review at the end of Chapter 1 to illustrate some techniques. These examples also point out the importance of correctly using parentheses in calculations.

Exercise 2: $\dfrac{98.20-98.60}{0.62}$

We will calculate the value in two ways. In doing so, we will intentionally make a mistake to show you how to correct errors using the [DEL] key. We will also discuss the Ans and Last Entry features.

Type 98.20 – 97..60 (an intentional mistake)

To correct the mistakes use the [◄] cursor key to move backward until your cursor is blinking on one of the double decimal points. Press [DEL] (on an 89, [♦][←] or position the cursor to the right of the character to be deleted and press [←]) and the duplicate decimal point will be deleted. Now press [◄] until the cursor is blinking on the 7. Type an 8, and it will replace the incorrect 7. On an 89, move the cursor to the right of the error, press [←] and then type the correct 8. Press [ENTER] for the numerator difference of ⁻4 as shown in the top of the screen below.

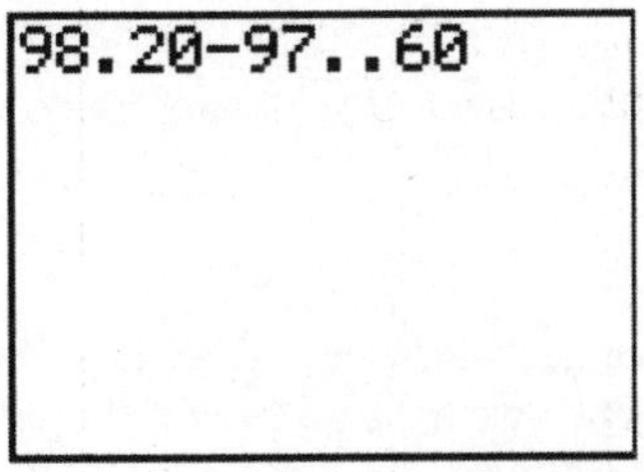

Press [÷]. (Note that "Ans/" appears on the screen). Type .62 and press [ENTER] for the result of ⁻.645.
Note: Ans represents the last result of a calculation which was displayed alone and right-justified on the Home screen. Pressing [÷] without first typing a value called for something to be divided, so Ans was supplied.

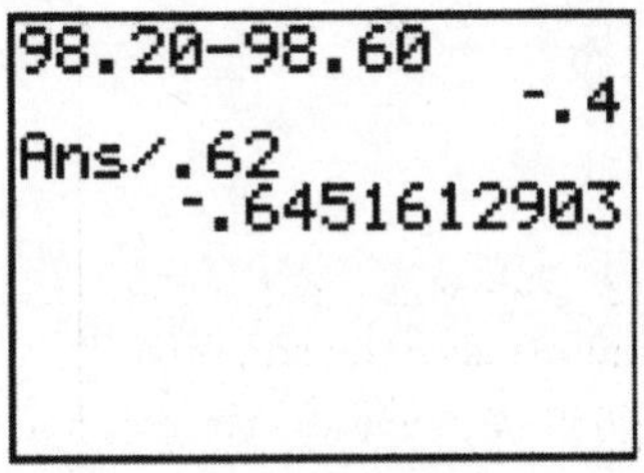

Press [2nd] [ENTER]. This calls the "last entry" to the screen. (in this case Ans/0.62). Press [2nd] [ENTER] again to get back to 98.20-98.60. Press the [▲] key to move to the front of the line. On an 89, press [2nd][◄].

On an 83 or 84, you will need to press [2nd] [DEL] (for INS); 89's are always in insert mode. You will see a blinking underline cursor. Type [(] to insert a left parenthesis before the first 9. Press [▼] ([2nd][►] on an 89) to jump to the end of the line. Type [)] [÷] 0.62 to see the result. Press [ENTER] for the same result as before.

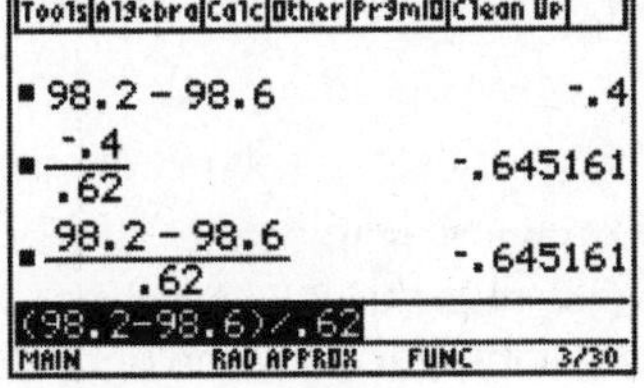

Exercise 5 $\sqrt{\dfrac{(5-7)^2+(12-7)^2+(4-7)^2}{3-1}}$

In this example, we will use the ANS function, illustrate syntax errors and show how to store quantities using variable names.

Type (5-7) ^{2}+(12-7)^2+(4-7)^2 as in the screen. Press [ENTER] for the value 38. (Use the [x^2] key for the ^{2})

```
(5-7)²+(12-7)²+(
4-7)²
                     38
√(Ans/(3-1)
            4.358898944
```

Press [2nd] [x^2] [2nd] [(-)] [÷] and then type (3-1). Press [ENTER] for the desired results at the bottom of screen (8). Note that the [2nd] [x^2] sequence is the $\sqrt{\ }$ function on an 84 or 84; on an 89 it is [2nd][×]. The [2nd] [(-)] sequence calls the last answer, Ans back to the screen.

In the screen at right, we attempted to do the whole exercise in one step.

```
√((5-7)²+(12-7)²
+(4-7)²)/(3-1))
```

Pressing [ENTER] brings this message because we have made an error. Press 2 to "goto" the error.

We get this screen which has a blinking cursor on the last parenthesis. This means we have an extra right parenthesis which has no matching left parenthesis.

This screen shows the result when we go back and insert the missing left parenthesis into the calculation. We get the same result as before.

The screen at right shows the same calculation done on an 89 calculator. One important difference here is that the 89 does not have the [x^2] key. To exponentiate to any power, use the [^] key followed by the desired power. Also notice the [▶] at the right of the output display. This is a cue that there is more to be seen. Press the up arrow to highlight the output display, then press to right arrow to scroll to the end.

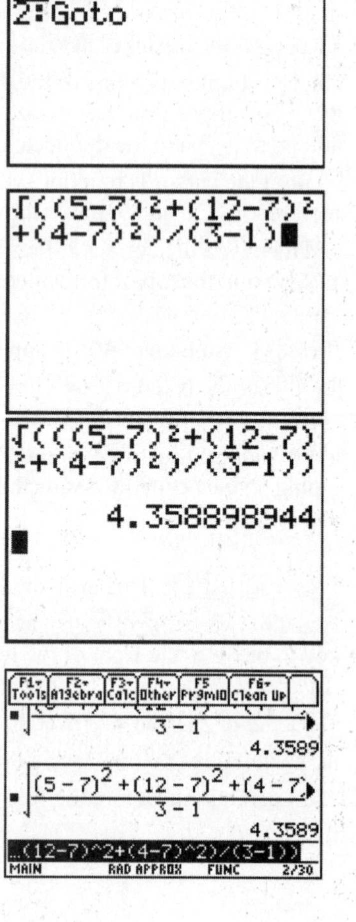

STORING LISTS OF DATA

Your text, like the real world, is full of sets of real data. In order to analyze the data, you must store it in a list in your calculator. Here, we will show you how to do this. Additionally, you should know that all of the data sets in the Appendix of your textbook are available to be transferred to your calculator from another calculator (or from a computer). You can then use the [APPS] key (on an 83 Plus or 84) to transfer a set into your lists. See your instructor and/or the Appendix of this companion for more information.

Storing Lists of Data from the Home Screen

EXERCISE: Random Sample of 10 Students: Revisited from pages 5 and 6 of this companion: Ten random integers between 1 and 60 were generated and displayed in a list. We will modify the example to store the integers in list L2 on a TI-83 or 84 (`list2` on an 89). We will then store the data in a list which we name ourselves.

```
222→rand
                222
randInt(1,60,10)
→L2
{31 1 2 11 27 5...
```

Modify the command by repeating the first three lines, but this time stipulate that the integers you generate will be stored in L2. Do this by typing [STO▸] [2nd] [2] after the `randInt`(1,60,10). Press [ENTER] for the screen at right. (Remember, that this was done explicitly in the random number generation on an 89).

```
222→rand
                222
randInt(1,60,10)
→L2
{31 1 2 11 27 5...
{31,1,2,11,27,52
,47,4,15,46}
```

To recall the list you have stored, so that it all appears on the Home screen press [2nd] [STO▸] [2nd] [2]. (The sequence [2nd] [STO▸] chooses the RCL (or recall) function and [2nd] [2] chooses list L2 as what will be recalled.)

```
F1▾   F2▾     F3▾  F4▾   F5     F6▾
Tools Algebra Calc Other PrgmIO Clean Up
RECALL VARIABLE
Recall: 
Enter=OK          ESC=CANCEL
MAIN     RAD APPROX    FUNC
```

To recall a list on the home screen of an 89, we also use the recall ([2nd] [STO▸]) command. However, this brings up an interim screen, which asks what you want to recall.

Press [2nd][-] ([VAR-LINK]) and arrow to highlight the list or variable you want. (Remember, for the TI-89 example, we stored the random numbers in `list1`.)

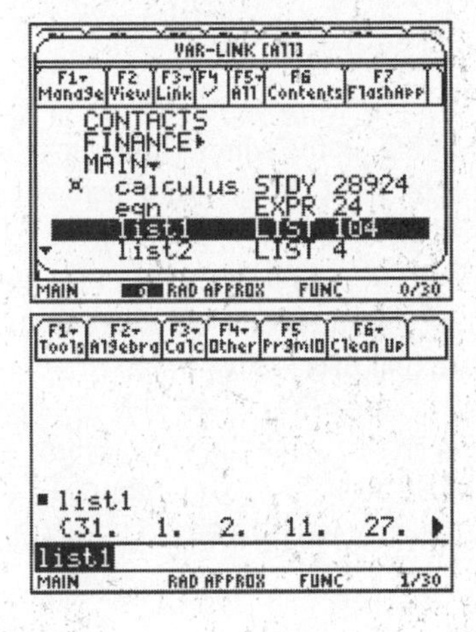

The name of the variable or list you want to display is echoed in the input area. Pressing [ENTER] recalls `list1`. Notice the ▸ again, which means you can up arrow to highlight the list, and move through it by pressing the right or left arrows.

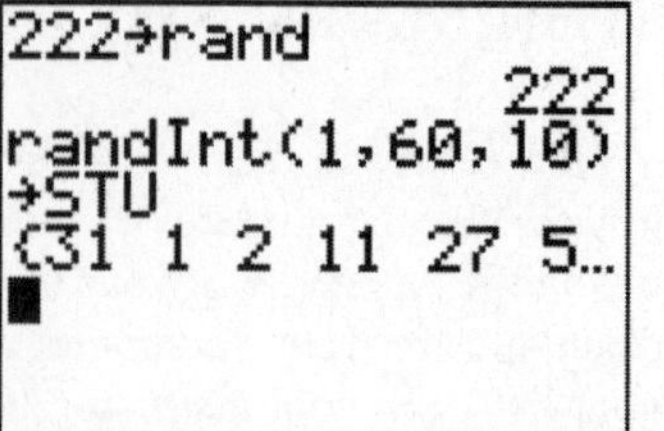

Now let's suppose we decided to store the list of random integers which correspond to students in a list named STU. We would simply modify the command so that instead of L2 we would store the data in a place called STU. Recall that you must use the [2nd][ALPHA] key to lock into alpha mode to type the name "STU".
Note: The TI calculator will know that STU is to be a list without being told because you are asking it to store several numbers, not just one.

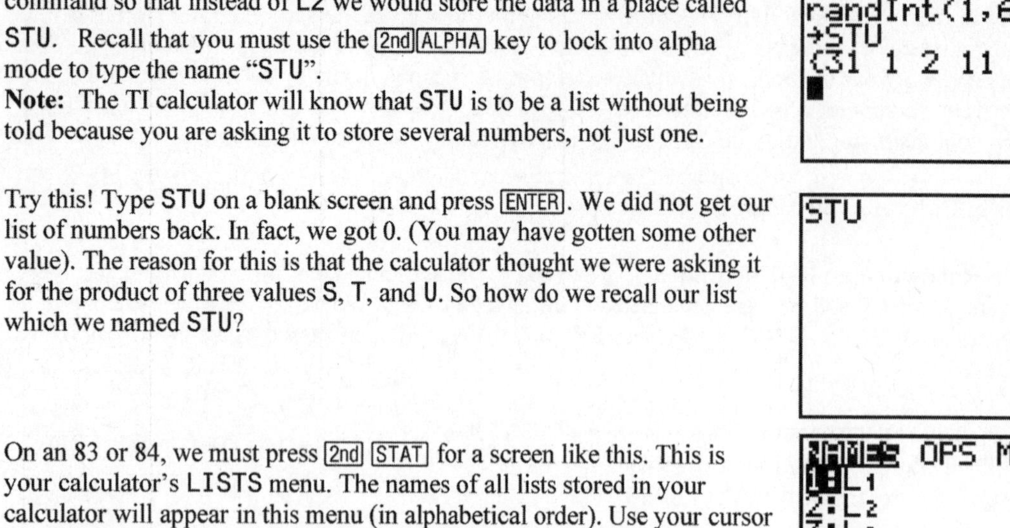

Try this! Type STU on a blank screen and press [ENTER]. We did not get our list of numbers back. In fact, we got 0. (You may have gotten some other value). The reason for this is that the calculator thought we were asking it for the product of three values S, T, and U. So how do we recall our list which we named STU?

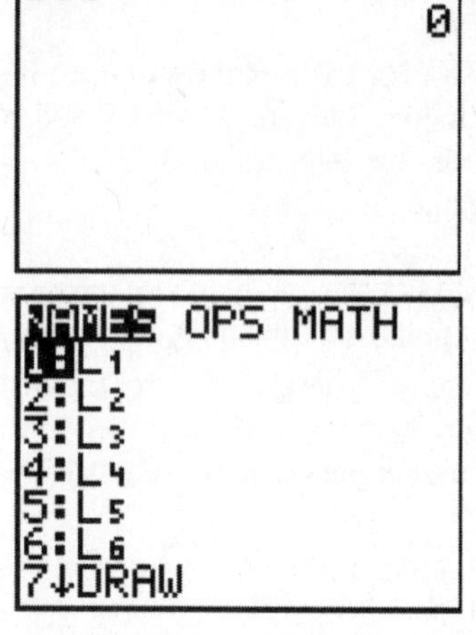

On an 83 or 84, we must press [2nd] [STAT] for a screen like this. This is your calculator's LISTS menu. The names of all lists stored in your calculator will appear in this menu (in alphabetical order). Use your cursor to look down the list of NAMES until you highlight the name STU.
Press [ENTER] twice.

ʟSTU
{31 1 2 11 27 5...

You should see this. You will notice that what was pasted on the Home screen is ʟSTU and not simply STU. So while we did not need to specify that STU was a list when we first stored our data, we did need to specify this fact when we were trying to recall it. In situations where it is unclear whether or not you need the small L in front of the name of a list, your safest bet is to paste the name in from the LISTS, NAMES menu as we did above.

Storing Lists of Data Using the STAT Editor

Using the STAT Editor is the easiest way to store lists and work with the data therein. The STAT Editor comes with 6 lists named L1 through L6 (list1 through list6 on an 89). Other lists can be added if desired. The number of lists is only limited by the memory size.

SetUpEditor

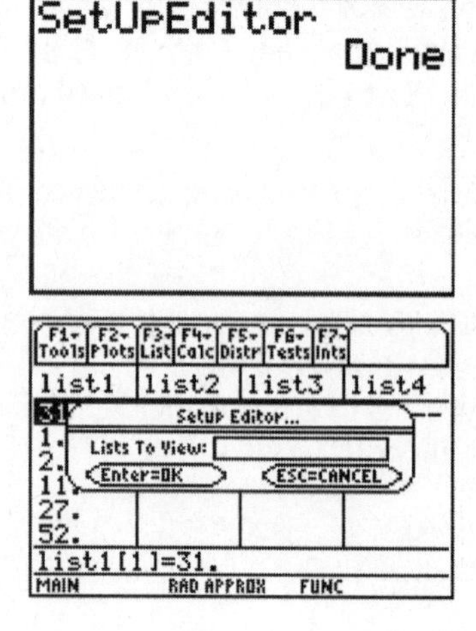

Setting up the editor will remove unwanted lists from view. It also will recover lists which have inadvertently been deleted
On an 83 or 84, if you want the STAT Editor to be restored to its original condition (with lists L1 to L6 only), press [STAT] [5] [ENTER]. Often students find this necessary because they have inadvertently deleted one of the original lists.

On an 89, in the Statistics Editor, press [F1] (Tools), then select option 3:Setup Editor. You will see the screen at right. Leave the box empty and press [ENTER] to return to the six default lists.

Clearing Lists in the Stat Editor

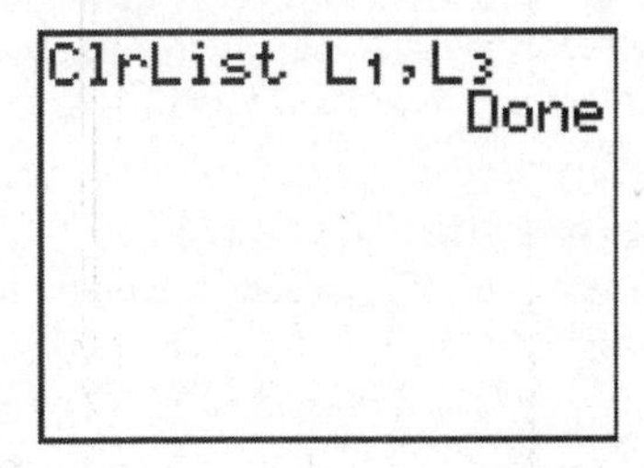

Let's say you want to clear out the contents from lists L1 and L3 before you start a problem. On an 83 or 84 (there is no command like this on an 89), press [STAT] [4] to choose the ClrList option from the Editor. This pastes the start of the command onto the Home screen. Then press [2nd] [1] [,] [2nd] [3] to choose your two lists with a comma between them as shown. The comma is needed if you are clearing multiple lists at one time as we are here. Press [ENTER] to see the message that the command has been Done.

View or Edit in the STAT Editor:

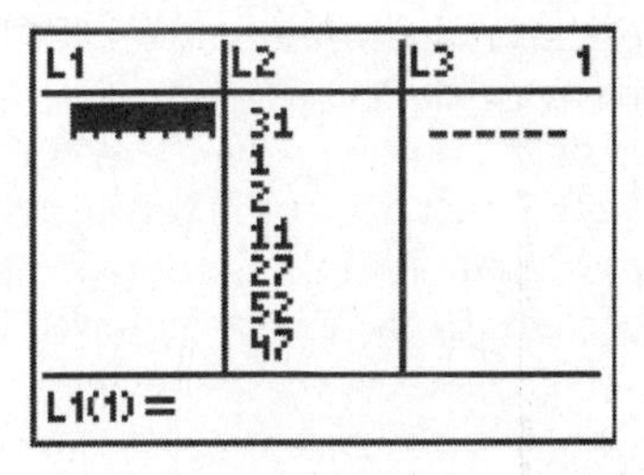

If you press [STAT] [1] on an 83 or 84, you should this screen with L1 and L3 cleared out and with our 10 random integers still stored in L2 (unless you have used the Statistics editor in some other class before). If you are using an 89, starting the Stats/List Editor app displays the editor.

Entering Data into the STAT Editor

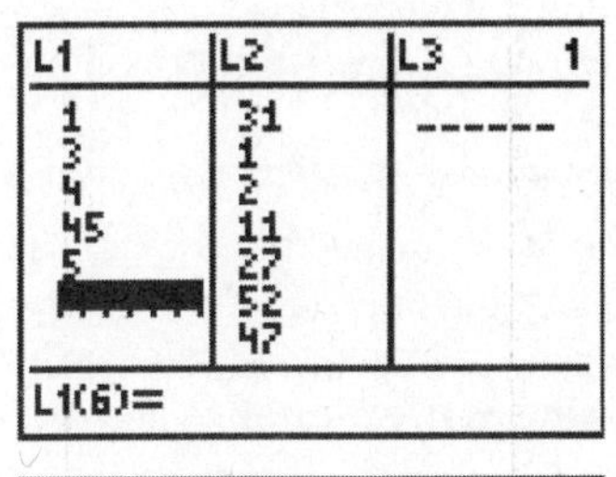

With the cursor at the first row of L1, type 1 and press [ENTER]. The cursor moves down one row. Type 3 followed by [ENTER] and the 3 will be pasted into the second row of the list. Continue with 4, 45, and 5 as seen at right.

Correcting Mistakes with DEL and INS

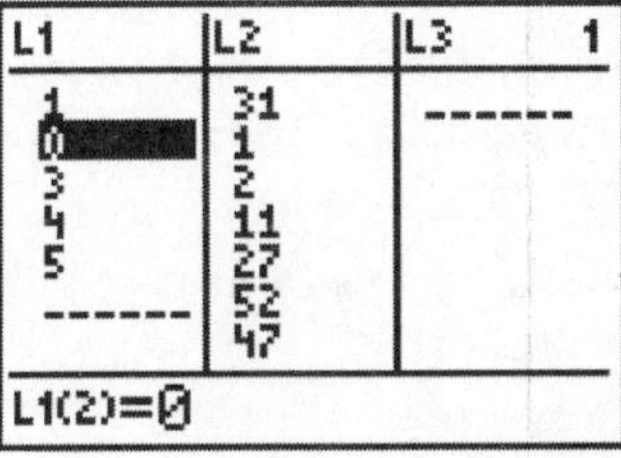

In the screen above, we can delete the 45 by using the [▲] key until it is highlighted and then pressing [DEL] ([♦][←] on an 89).

To insert a 2 above the 3 move the cursor to the 3 then press [2nd] [DEL] ([2nd][←] on an 89) (to choose INS or insert mode). Note a 0 was inserted where you wanted the 2 to go.

Clearing Lists without Leaving the STAT Editor

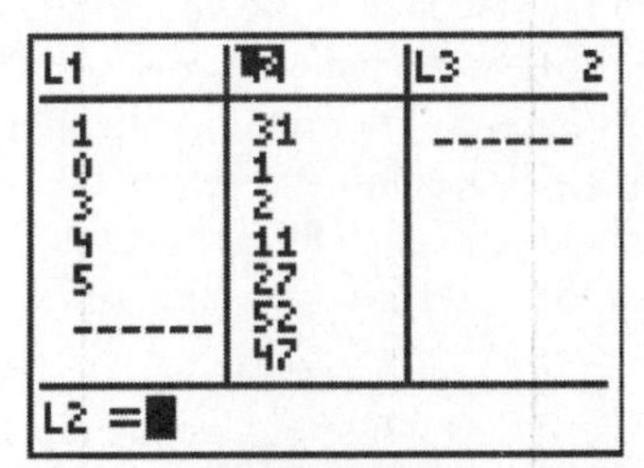

Suppose you wish to clear a list, say L2, while you are still in the STAT Editor. You should use the cursor to highlight the name of the list at the top. With the name highlighted press [CLEAR] and you will see this. Press [ENTER] and the contents of the list will be cleared. *Make sure not to press* [DEL] or the list will be deleted entirely and you will have to use SetUpEditor as described above to retrieve it.

Storing Data with a Named List

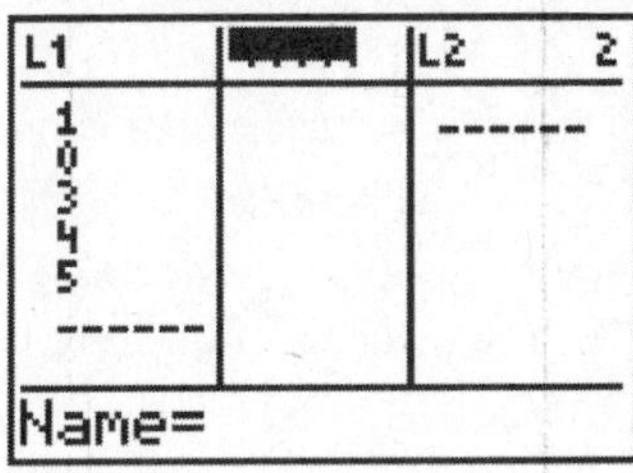

Suppose we wish to store the 10 random numbers 17, 44, 43, 28, 27, 51, 30, 39, 34, and 32 in a list named Rand2. This procedure is the same on all calculators.

First with L2 highlighted at the top press [2nd] [DEL]. L2 will move to the right and there will be a new list inserted in its place. The calculator will be in ALPHA mode and be prompting you for the name of your new list.

Type RAND and then press ALPHA to release the calculator from ALPHA mode, so you can type the final character in the name which is 2.

Press ENTER and then ▼ (no need to press ⊙ on an 89). Now input the desired values into the list as at right.
Note: If the list Rand2 had already been created, its name could have been pasted into list name. In either case, if the list already existed its entire contents (if there were any) would have been pasted in as well as its name.

Deleting a List from the STAT Editor:

If you wish to delete a list from your STAT Editor, simply highlight the list name and press DEL. The name and the data are gone from the Editor but not from the memory.

Using SetUp Editor to Name a List:

On a TI-83 or -84 home screen, press STAT 5 to call SetUpEditor. Now type IDS, STU, L1. You will have to be careful to keep pressing the ALPHA key before each character or release the alpha lock (2nd ALPHA) to type the commas. Then press ENTER. Then press STAT 1 to view your lists.

In the screen at right, you will see the old list STU has been placed back in full view. Also there is a new list IDS ready and waiting for some data to be entered. Try pressing the right arrow to see more lists – there aren't any. This command set the editor with just the three lists specified.
To do this on an 89, follow the steps outlined above to generally set up the editor, and type in the list names just as one does for the 83 or 84.

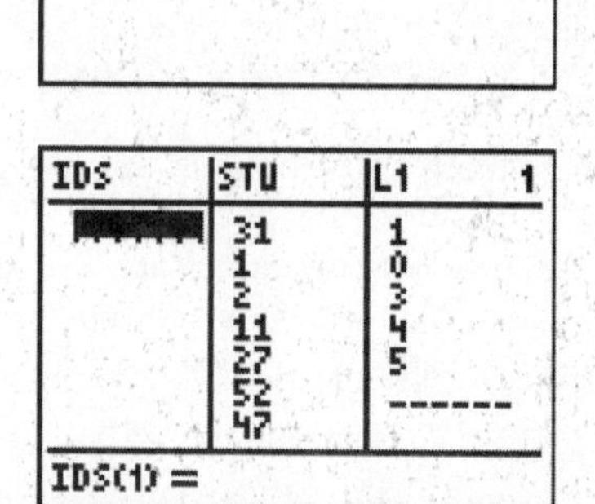

Making a Copy of a List

On an 83 or 84, use cursor control keys to highlight the top of the new list IDS as in my screen. Press 2nd 1 to paste in the name L1 as in the bottom line of the screen. Press ENTER and the L1 data will appear under IDS. The contents of L1 have been copied to IDS.

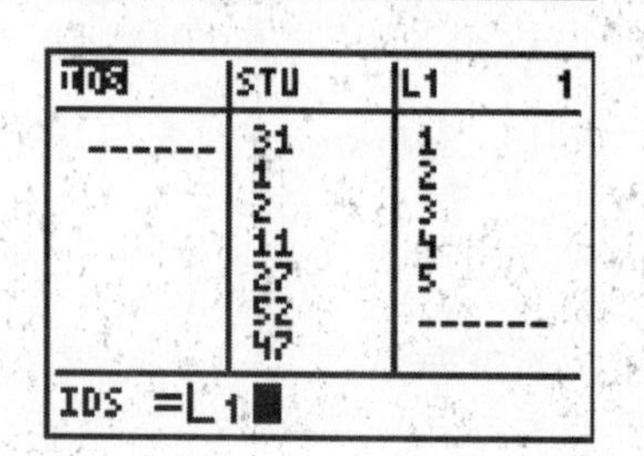

On an 89, use the cursor keys to highlight the top of the new list `ids` as with the 83,but press [F3] (`list`). Now, press [ENTER] to select option `1:Names`. Scroll down to find the name of the list you wish to copy, and press [ENTER] to select it. The list name will be pasted in as at right. Pressing [ENTER] again will copy the list.

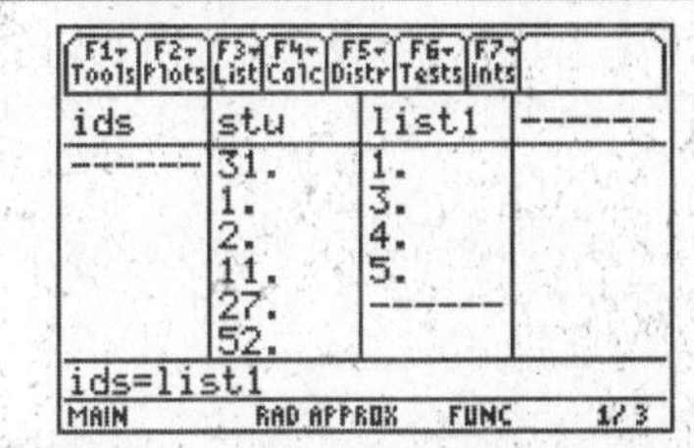

Generating a Sequence of Numbers in a List

From time to time one may want to enter a list of sequenced values (years for example in making a time-series plot). It is certainly possible (but tedious) to enter the entire sequence just as one would enter normal data. There is an easier option, however.

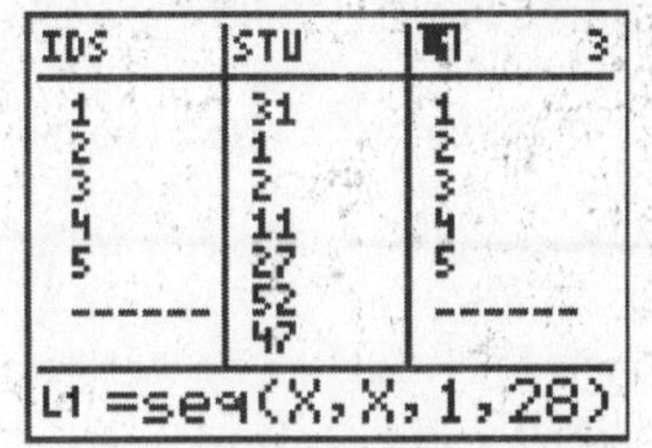

Use cursor control keys to highlight `L1` in the top line. Press [2nd] [STAT] [▸] [5]. You are choosing the `LIST` menu and then choosing the `OPS` submenu. From the `OPS` submenu you are choosing option 5 which is `seq(`. This has been pasted onto the bottom line of the screen. Type in the rest so that you have `seq(X,X,1,28`. Press [ENTER] and the sequence of integers from 1 to 28 will be pasted into `L1` as in my screen.

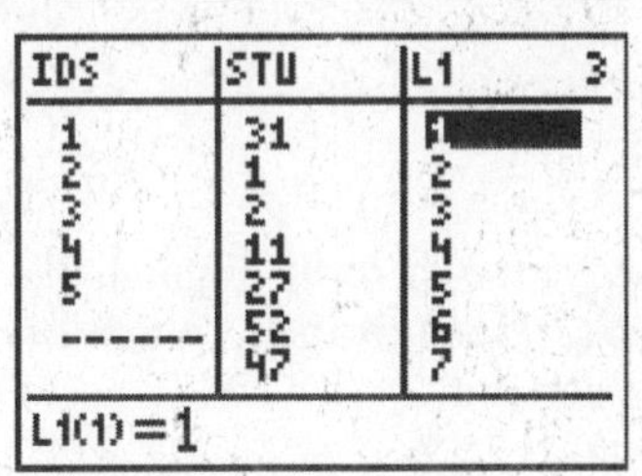

On an 89, the procedure is analogous, but access the `LIST OPS` menu by pressing [F3][2], then select option 5.

Note: To quickly check the values on a multi-screen list you can press the green [ALPHA] key followed by either the [▴] or [▾] key. This will allow you to jump up or down from one page (screen) to another. The green arrows on the keyboard near the [▴] and [▾] keys are there to remind you of this capability. On a TI-89, instead of the [ALPHA] key, press [◆].

DELETING LISTS

Deleting a Named List from Memory (TI-83/84)

To remove both the name of a list and its data from RAM press [2nd] [+] to call the `MEM` function. You should see this screen.

Press [2] to choose `Mem Mgmt/Del`. You will get a screen like this.

Press [4] to choose to see a list of *List* names. Use [▼] to move down the display to the list you want to remove (say `IDS`) as shown in this screen.

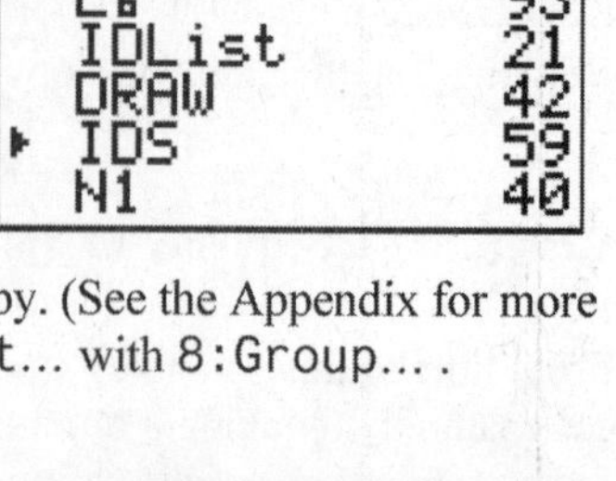

Press [DEL] to delete the list. You can remove lists one by one from this screen.

Press [2nd] [MODE] to `QUIT` and return to the Home screen.
Note: (For TI-83 Plus and TI-84 only) If the list was also saved in a Group stored in Archive memory, the above procedure has not deleted the copy. (See the Appendix for more information on Groups.) Groups can be deleted from the calculator if `4:List...` with `8:Group...` .

Deleting a Named List from Memory (TI-89):

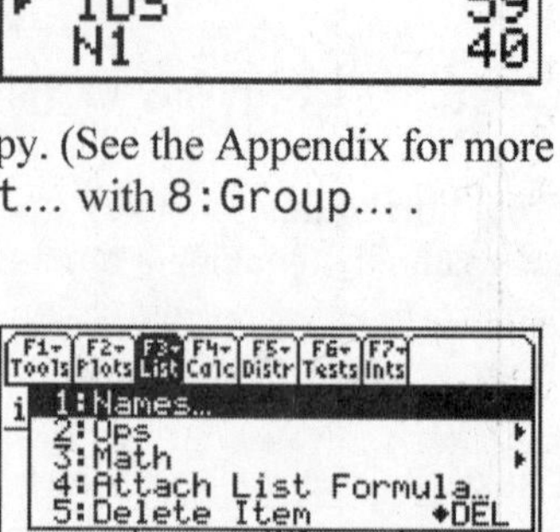

In the Statistics/List Editor, Press [F3] (`Lists`) and press [ENTER] to select option `1:Names`.

Move the cursor to highlight the name of the list you want to delete. Now, press [F1] (`Manage`). Option 1 on this submenu is `Delete`. Press [ENTER] to select it.

You will get the screen at the right which asks you to confirm the deletion. Pressing [ENTER] will complete the delete, [ESC] will cancel the operation.

Using the Augment Function for Large Lists

Several lists can be combined into one large list by using an option which is on the `LIST, OPS` submenu at option 9. This is the augment option. For example if you wish to combine lists `L1` and `L2` and store the combined list in `L3`, you can press [2nd] [STAT] [►][9] to paste the `augment(` on your Home screen. Then put the names of the lists you wish to combine on the screen separated by commas. Close the parentheses. Then press [STO►][2nd] [3] to store in `L3`. Finally press [ENTER]. This function will only work if there is actually data in the lists you are augmenting.

To do this procedure on an 89, place the cursor at the top of an empty list to highlight the list name, then press [F3] (`List`). Press [2] to select the `Ops` menu, where the augment function is choice 8. Press [8] to paste it into the input area. Now press [F3] again followed by [ENTER] to select the names list. Arrow to the names of the lists you want to combine and press [ENTER] to paste their names into the command. Be sure to separate each name with commas and finish the command by closing the parentheses.

Note: The "augment" function could be used so that several students could cooperate in typing in a large set of data. Each student could type part of the data set into a different list (`L1`, `L2`, `L3`, etc.) Then the calculators could be linked, the lists transferred to one calculator and then the augment function could be used to combine them into one large list which could then be redistributed for all to share.

WHAT CAN GO WRONG?

Why is my list missing?

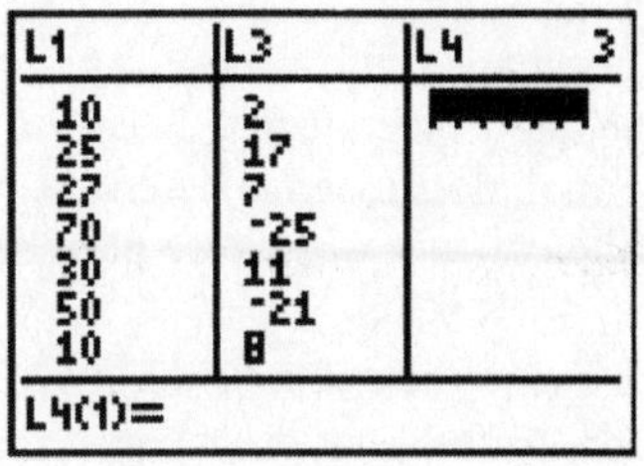

By far the most common error, aside from typographical errors is improper deletion of lists. When lists seem to be "missing" the user has pressed [DEL] rather than [CLEAR] in attempting to erase a list. Believe it or not, the data and the list are still in memory. To reclaim the missing list press [STAT] and select choice `5:SetUpEditor` followed by [ENTER] to execute the command. Upon return to the Editor, the missing list will be displayed.

L1	L3	L4 3
10	2	
25	17	
27	7	
70	-25	
30	11	
50	-21	
10	8	

L4(1)=

2 Summarizing and Graphing Data

This chapter introduces the graphical plotting and summary statistics capabilities of the TI calculators. These calculators can display histograms, scatter (*X-Y*) plots, connected scatter plots (used for time-series graphs), normal plots (used to determine whether a data set is approximately bell-shaped), and boxplots. This chapter will deal only with those plots described in Chapter 2 of the text.

First row keys like [2nd] [Y=] (STAT PLOTS) on the TI-83 and -84 are used to obtain descriptive plots of data sets. On the TI-89, plots must first be defined and then displayed. As the procedures are somewhat different due to the differences in menus, we will first describe the procedures for the 83/84 calculators, then repeat as necessary the discussion for the 89 series. The author assumes you have read and familiarized yourself with the content of Chapter 1.

We will use the data on the ages of Best Actress Oscar winners presented in Table 2-1 of the text to illustrate first the procedure for creating a histogram from raw data. This is by far the most common means of displaying data using technology. Procedures for using data which have already been summarized into a table will be presented later. The table is repeated below for convenience.

Ages of Best Actress Winners									
22	37	28	63	32	26	31	27	27	28
30	26	29	24	38	25	29	41	30	35
35	33	29	38	54	24	25	46	41	28
40	39	29	27	31	38	29	25	35	60
43	35	34	34	27	37	42	41	36	32
41	33	31	74	33	50	38	61	21	41
26	80	42	29	33	35	45	49	39	34
26	25	33	35	35	28				

Enter the ages into list L1 (TI-83/84) or list1 (TI-89).

Plot1 Plot2 Plot3
\Y1=
\Y2=
\Y3=
\Y4=
\Y5=
\Y6=
\Y7=

One preparatory step which should be undertaken before defining any statistics plot is to ensure that all functions entered on the [Y=] screen have been erased; if not, these will also display and may also cause dimensioning problems for the calculator. Press the [Y=] key on the upper right on a TI-83/84, or go to the Y= app on a TI-89 and ensure that all functions are cleared (Press [CLEAR]) as at right.

CREATING A HISTOGRAM – TI-83/84

The next step is to define the plot. This is done by pressing [2nd][Y=] (Stat Plot). You will see the screen at right. Notice that there are three plots which can be displayed at any one time. For most purposes, there should be only one turned "on" at once. Notice here Plot1 is On and Plots 2 and 3

show Off. Scrolling down the menu are options 4 and 5 that turn all plots off or on with a single command. Selecting these will transfer the command to the home screen. Executing it requires pressing [ENTER].

Press [ENTER] to select Plot1. The cursor should be blinking over the word On. If On is not already highlighted, press [ENTER] to move the highlight and to select displaying the plot. Notice there are six graphics types. Histograms are the third choice. Pressing [▼] will move the cursor to the first plot type. Use [▶] to move the cursor to the histogram figure. Press [ENTER] to move the highlight.

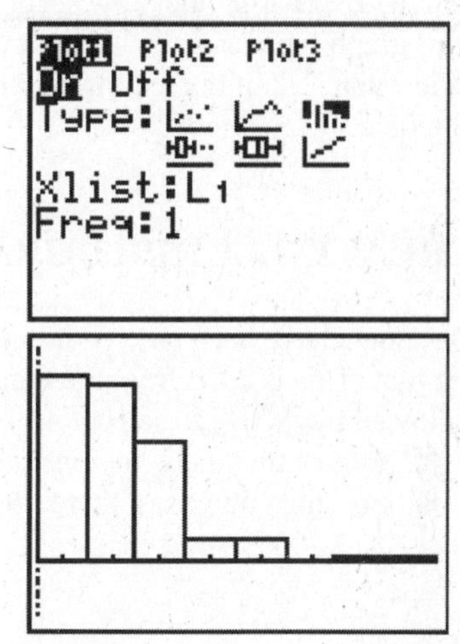

At this point, your screen probably looks like the one at right. We're ready to display the graph, since our data was in list L1 and each data value had frequency 1 (representing one occurrence of the value.) If you want to graph data in other lists, move the cursor to Xlist: and enter the list name ([2nd] n, where n is the number of the list). We'll talk more about frequencies later. Notice if you move the cursor to Freq: it will flash as [A]. If you need to change this back from something else to a 1 you will need to press [ALPHA] before typing the 1.

The easiest way to display a histogram (or any statistics plot) is to press [ZOOM][9] (Zoom Stat). The resulting graph is seen at right. The *X*-axis "floats" a little way up from the bottom of the screen. This is so that values as seen in the next picture do not interfere with the plot.

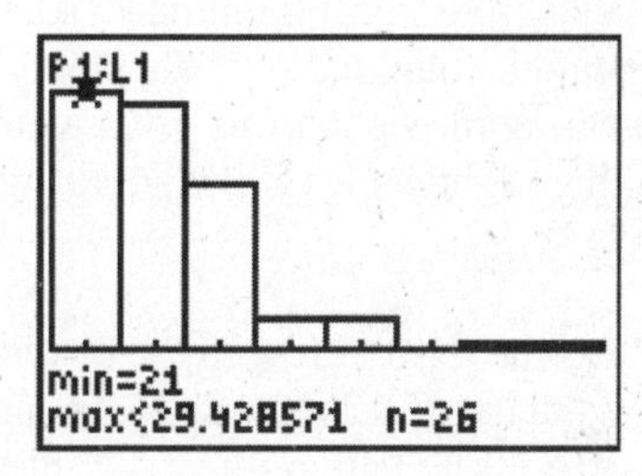

You will want to see exactly what the graph shows. To do this, press [TRACE]. A blinking cursor will appear in the first bar at the left of the graph. At the bottom of the screen the minimum value included in the bar, maximum value for the bar, and number of observations in the bar will be displayed. This bar goes from 21 to 29.428571. There are 26 observations in this interval. Pressing the right arrow key ([▶]) will allow you to continue through the graph seeing the interval ranges and numbers of observations in each bar.

At this point, we can see the distribution of Best Actress winners' ages appears to be unimodal and right skewed, as the highest bar is on the left and frequencies drop consistently as we move to the right. Visually, the center of the graph (the midpoint) will be somewhere around 35 years old.

There is a downfall to using simply [ZOOM][9] for histograms. Look at the first interval. It doesn't really make sense in a natural way. The bar width represents a difference between the low and high ends of each bar of 8.428… which is unnatural. We'd like to fix this.

Manipulating Windows on the TI-83/84

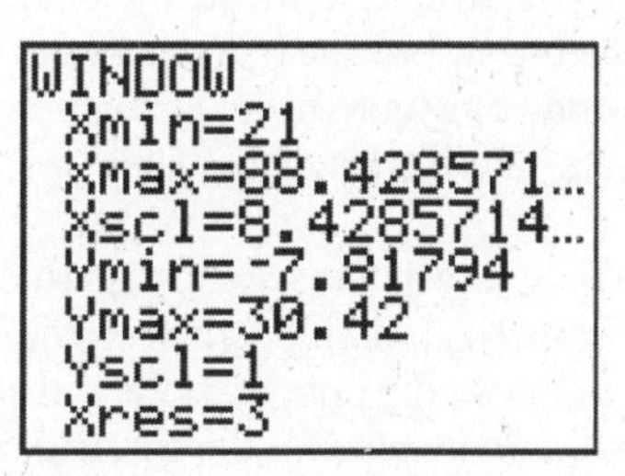

To force particular minimums, maximums and scaling we will press [WINDOW]. This displays the screen at right. Notice the Xmin was the smallest value shown on the plot and Xscl was the bar width. These are what we'd like to change. You generally won't have to change any of the *Y* variables here (unless you lose the top of a bar – then increase Ymax).

Change Xmin to 21 and Xscl to 10 (This sounds pretty reasonable and will duplicate the frequency distributions used in the text, since all ages are integers and < 31 is really 30 or less).

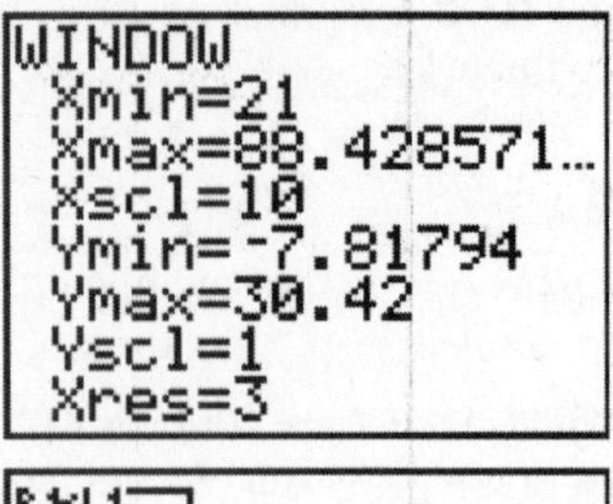

To display the new graph, press [GRAPH]. NEVER press [ZOOM][9] after changing a window. You'll just go back to the one you had before! Notice that the first bar includes ages from 21 to less than 31 (or at most 30). This graph replicates the histogram one would get using the tabulated data as in Table 2-2 of the text, from which the histogram in Figure 2-2 was obtained.

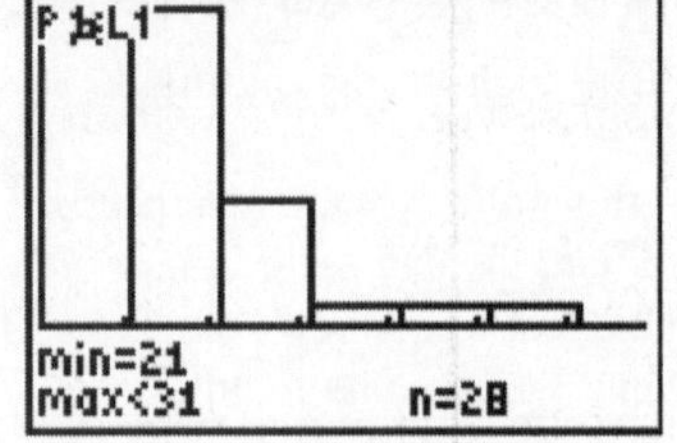

CREATING A HISTOGRAM – TI-89

Once data have been entered into a statistics list, the first step is to define the plot. This is done from the Statistics Editor by pressing [F2] (Plots) followed by [ENTER] to select Plot Setup. You will see the screen at right. Notice that there are nine plots which can be displayed at any one time. For most purposes, there should be only one active (checked) plot at once.

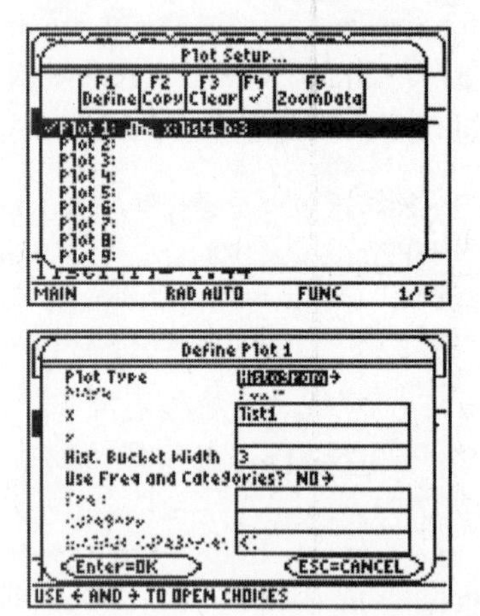

Press [F1] to select defining Plot1 since it was highlighted. The cursor should be blinking over the plot type. If not already set to a histogram, pressing the right arrow gives a menu of five plot types. Move the cursor to highlight choice 4:Histogram and press [ENTER] to select it or press [4].

Press the down arrow to the box labeled x. Press [2nd][-] (VAR-LINK) to get the list of list names. Move the cursor to highlight the one you wish to use, press [ENTER] to select it. The TI-89 then wants the histogram bucket width which is the bar width. Press the down arrow to move to this box. This is something that may have to be "played around with" to get a good picture. Here, I have set the bar width to 10. Press [ENTER] to complete the plot definition. You will be returned to the Plot Setup menu.

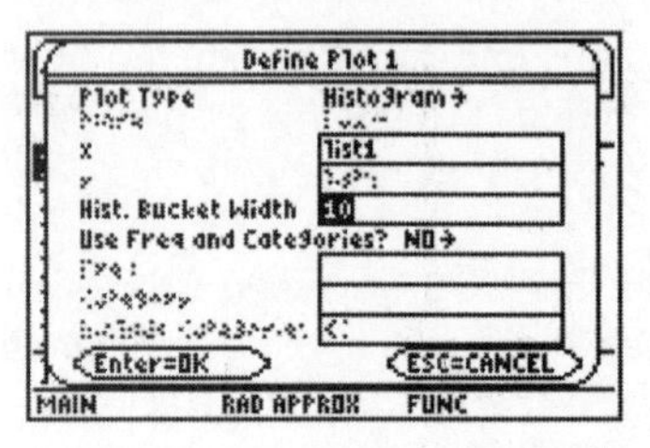

The easiest way to start displaying a histogram (or any statistics plot) is to press [F5] (Zoom Data). The resulting graph is seen at right. The x-axis "floats" a little way up from the bottom of the screen. This is so that values, as seen in the next picture, do not interfere with the plot. This picture shows one bad point of the TI-89. It does not always get the windowing correct for histograms. We don't see the full heights of the bars. We'll change that later.

You will want to see exactly what the graph shows. To do this, press [F3] (Trace). A blinking cursor will show in the first bar at the left of the graph. At the bottom of the screen the minimum value included in the bar, maximum value for the bar, and number of observations in the bar will be displayed. This bar goes from 15.1 to 25.1. There are seven observations

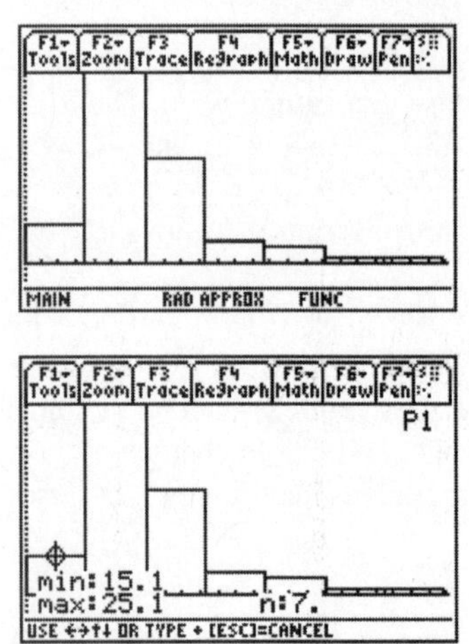

in this interval, as indicated by the `n:7` at the bottom right. Pressing the right arrow key (▶) will allow you to continue through the graph seeing the interval ranges and numbers of observations in each interval.

At this point, we can see the distribution of winning actresses ages appears to be unimodal and relatively right-skewed as distribution is much longer on the right-hand side of the peak. We see the center is around an age of 35. We'd really like to see the whole graph, however.

There is another downfall to using simply `Zoom Data` for histograms. Look at the intervals. They really do not make sense in a natural way. We'd like to fix this.

Manipulating Windows on the TI-89

To force particular minimums, maximums and scaling we will press [◆][F2] (`Window`). This displays the screen at right. Notice the `xmin` was the smallest value shown on the plot; `xmax` is the largest. `ymin` and `ymax` are analogous. `xscl` and `yscl` are the distances between axis "tick marks".

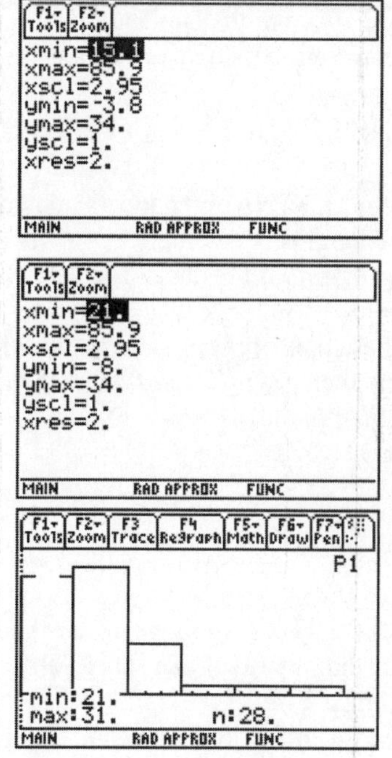

Change `xmin` to 21 (the smallest data value) `ymin` to –8 and `ymax` to 34. Why must `ymax` be so large? The "tabs" at the top of the screen will hide some of the plot unless it is sized on the roomy side. `Ymin` is set low so that the legends which appear after pressing `Trace` don't obscure portions of the graph.

To display the new graph, press [◆][F3] (`Graph`). This looks better, and also replicates the histogram one would get using the tabulated data as in Table 2-2 of the text, from which the histogram in Figure 2-2 was obtained.

CHANGING THE NUMBER OF BARS

Histograms are a subjective type of plot. How many bars to display (and what the lowest display value for the variable) can be manipulated at will until one has a "nice" plot. Your instructor may give you some guidelines. One "rule of thumb" for many years was to have somewhere between 5 and 20 bars; for most smaller data sets dividing the number of observations by 5 gives a good estimate of how many bars will give a decent picture.

On TI-83/84 calculators, the bar width is controlled using `Xscl` on the window settings screen. On theTI-89 series, one changes the bucket width on the plot setup screen.

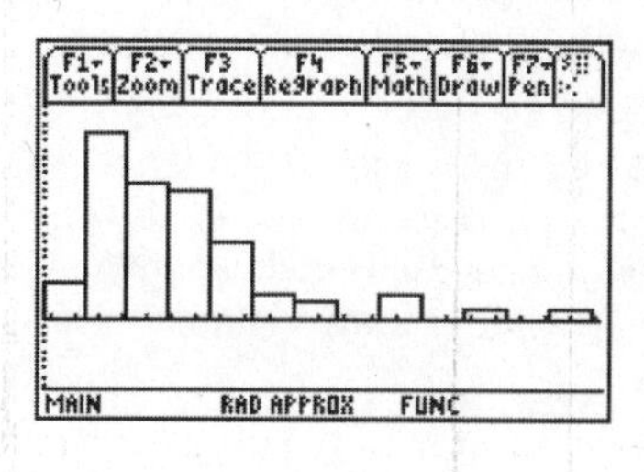

The screen at right was created using an `Xmin` of 20 and a bucket width of 5. `Ymax` was reset to 25 for picture resolution. Just as having too few intervals can hide the true nature of a distribution, having too many can also distort the picture. But here again, we see the unimodal, right-skewed distribution of the previous histogram.

"Printing" the Picture

Unfortunately, calculators do not have printers. To make a hard copy of the graph once you are satisfied with it, use the [TRACE] key to examine the entire graph. Draw a picture of the histogram, clearly labeling each axis and giving the graph a title. Remember that the intervals given are the endpoints of the intervals. Label them as such. When you are finished, you should have a picture like the one at right. If you have TI-Connect software, you can use the screen capture application to save the picture on the computer for printing directly from the application, or include the graph as a part of a word processing document.

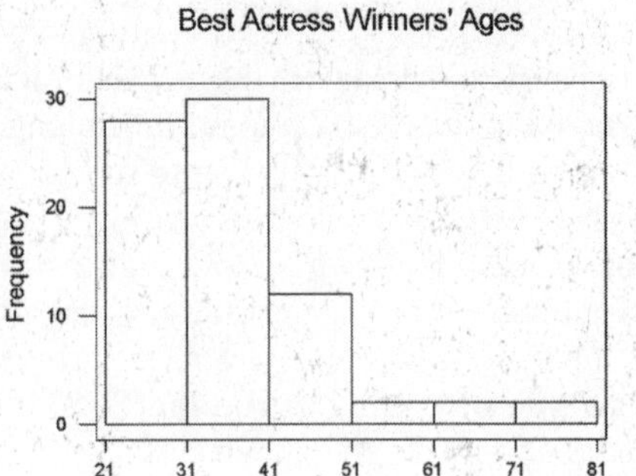

HISTOGRAMS FOR TABULATED DATA

When data are given in a tabulated list, such as in Table 2-5 which gives the heights of a sample of 1000 women, we can still create a visual histogram of the data. The data are reproduced at right for convenience.

Height (in)	Frequency
56-57.9	10
58-59.9	64
60-61.9	178
62-63.9	324
64-65.9	251
66-67.9	135
68-69.9	31
70-71.9	6

These data are presented in two-inch intervals (the first interval is 56" to less than 58"). We will enter the midpoint of each interval (57, 59, 61, etc) into a list (L1 or list1) and the frequencies into a second list (L2 or list2).

On the TI-83 or 84, we define the plot as has been done before, but instead of using Freq:1 we specify the list containing the frequencies (L2). If using a TI-89, on the plot definition screen, use a bucket width of 2, and use the arrow key to change the answer to Use Freq and Categories? from No to Yes, then use Var-Link ([2nd][-]) to select the list with the frequencies.

Since we know the bar widths we need (2 inches) and the minimum and maximum data values, as well as the frequencies, it is easier in this case to specify the window settings as at right. Set the window for the TI-89 similarly, but remember to give extra height to the y-axis for room for the tabs.

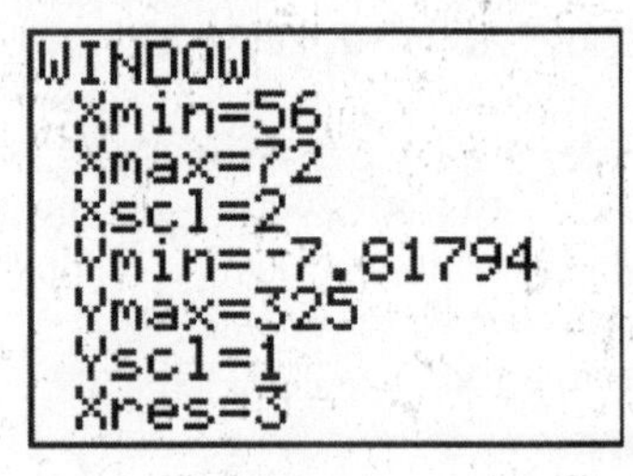

Here is the finished histogram. Compare it to the graphic produced by STATDISK shown in the text in section 2-3.

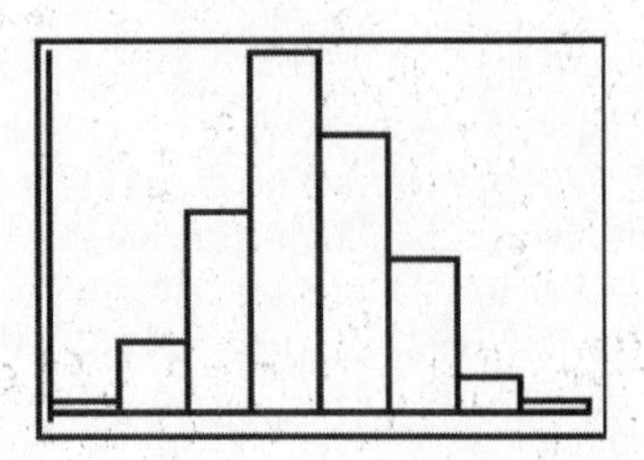

FREQUENCY POLYGONS

Ages	Female Frequency	Male Frequency
21 - <31	28	3
31 - <41	30	25
41 - <42	12	30
51 - < 61	2	14
61 - <71	2	3
71 - < 81	2	1

A frequency polygon is another type of plot which is used to picture a distribution. It uses line segments to join points that are located at the class midpoint on the *X*-axis and the frequency of the interval on the *Y*-axis. This type of plot is useful for comparing two distributions, as well. We return to the data on Oscar winners ages, reproducing for convenience the data in tabular form. I have labeled the age columns as extending to less than the beginning of the next interval to emphasize that these are, for practical purposes, 10 year intervals. This frequency chart clearly indicates that males tend to win Best Actor at ages older than females – a fact hidden in the long list of numbers in Table 2-1.

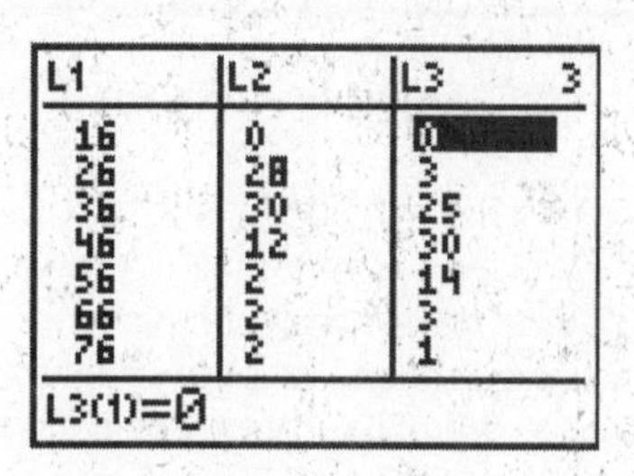

The midpoint of each interval is the average of the two ends, so the midpoint of each interval ends in 6. When entering the data in the calculator, we use one list for the age midpoints, and two lists for the frequencies. We will also insert two additional intervals (midpoints of 16 and 86) both with frequencies of 0. At right is the first screen of data.

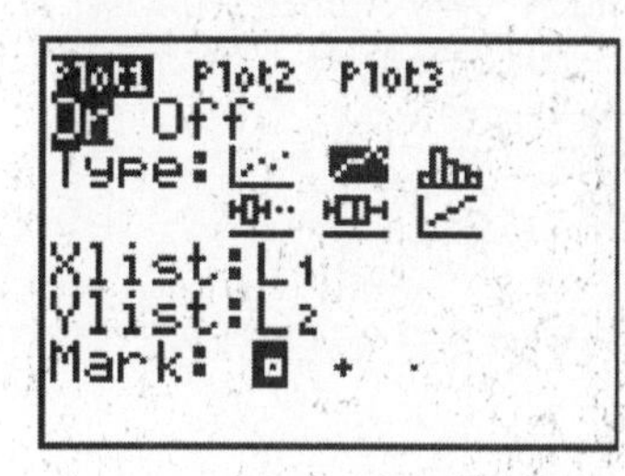

To display both distributions at the same time, we will define two graphs to display together. On the TI-83 or 84 we will use the second plot type (⊾). On the 89, use the arrow keys on the plot definition screen to select plot type `2:xyline`. The `Xlist(x)` is the one with the age midpoints (`L1` or `list1`) and the Ylist (y) is the one with the female frequencies. Choose a mark to display the actual data values. My TI-84 plot definition is at right.

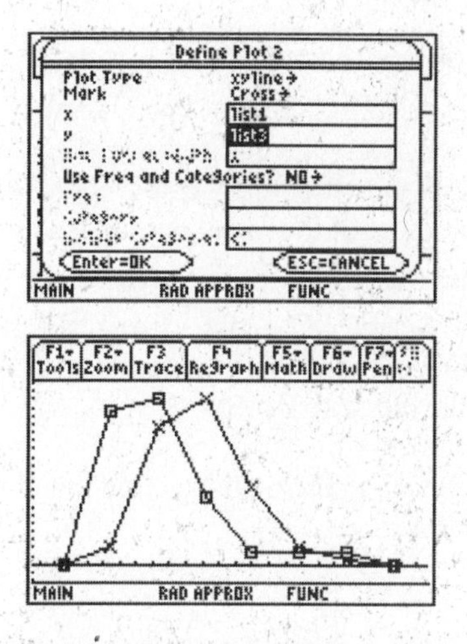

Now we will define a second plot similar to that just done, but using the frequencies (Ylist) for the male winners. Select a different type of mark for this plot to help distinguish the two. My definition of the second plot on the TI-89 is at right.

On both calculators, `Zoom Data` ([ZOOM][9] or [F5]) will display the graph. The graph with the square points is for women, the crosses are for the male data. We clearly see again a difference in the age distributions – the men's distribution seems to resemble the women's, but shifted older by 10 years .

Ogives

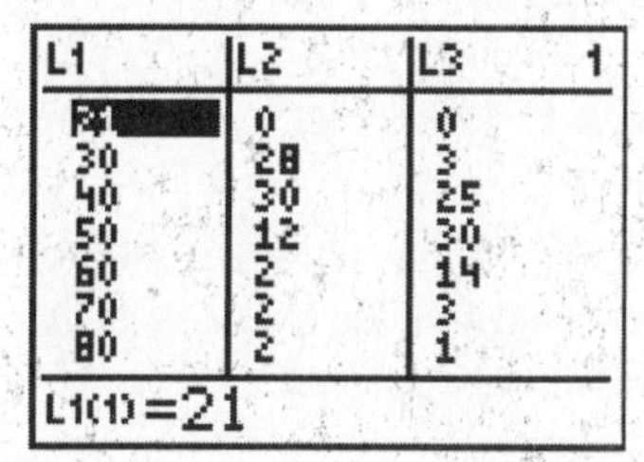

Ogives are a cumulative frequency polygon. Like the frequency polygons above, they are also connected line plots, but they use the category end values instead of midpoints on the *X*-axis. Since the lowest age of the Oscar winners was 21, and practically speaking, the largest age in each category is 30, 40, etc I have changed list L1 to reflect these boundaries. Ogives are most commonly displayed with percents (relative frequencies) on the *Y*-axis, although one can also use actual frequencies.

We need to construct the cumulative frequency distribution(s). In the Statistics List editor, move the cursor to highlight the next empty list (L4 or list4 on the 89). On both calculators, we want to do list arithmetic, so on the 83/84 press [2nd][STAT] (LIST) on the 89 press [F3]. Now arrow to OPS and all calculators select choice 6:cumsum(. Enter the list name you want to sum (L2 or list2), close the parentheses then type /76*100 to divide by the total number of winners, and then convert the decimal to a percentage. Press [ENTER] to complete the calculation.

list1	list2	list3	list4
16.	0.	0.	------
26.	28.	3.	
36.	30.	25.	
46.	12.	30.	
56.	2.	14.	
66.	2.	3.	

list4=...Sum(list2)/76*100^

As with frequency polygons, on both calculators Zoom Data ([ZOOM][9] or [F5]) will display the graph. With [TRACE] activated, we see that 76% of all female Oscar winners were at most 40 years old.

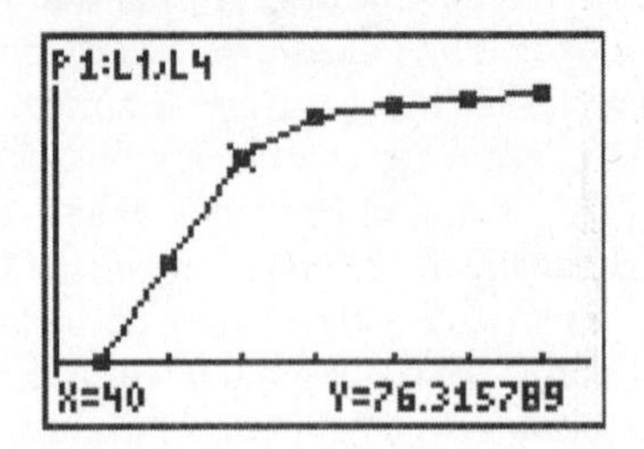

In the screen at right, I have repeated the process as described above using the male winner's data and defined a second plot as we did for the frequency polygons. Here we also see the dramatic right shift, at least until at ages of 60 and older, both genders show similar behavior.

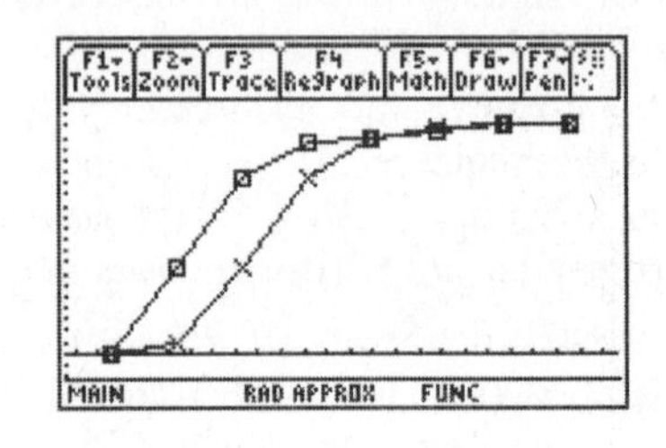

DOTPLOTS AND STEMPLOTS

TI calculators will not create these types of plots. However, sorting the raw data can help in creating them "by hand."

On the TI-84/84 series calculators, the Sort command is located on the initial [STAT] menu. There are two choices: 2:SortA(and 3:SortD(which sort the data in ascending (low to high) or descending order. In either case, select the appropriate choice by either pressing the desired number of the menu option or by arrowing to move the highlight and pressing [ENTER]. The command will be transferred to the home screen. Enter the name of the list to be sorted and finally press [ENTER] to execute the command. The calculator will tell you when it is finished by showing the message Done.

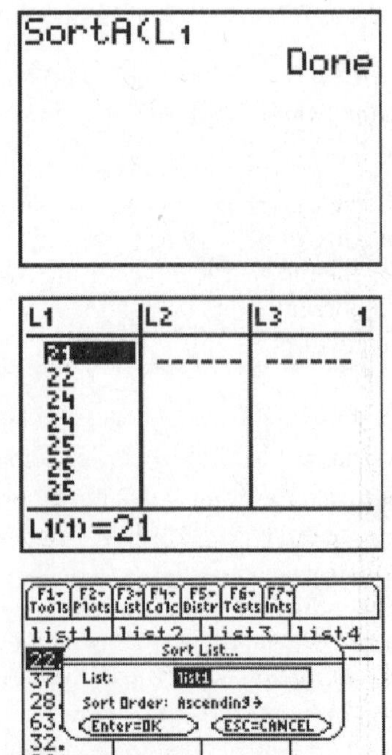

Return to the Statistics Editor and you can then page through the sorted list of data.

On a TI-89 calculator, within the Statistics List Editor, select the List menu ([F3]) and arrow to Ops. Pressing ⊙ displays the menu. Press [ENTER] to select choice 1:Sort List. You enter the name of the list to be sorted and determine ascending or descending order. Pressing [ENTER] will execute the sort.

PARETO CHARTS

Pareto charts are a special bar graph for categorical data, where the bars are sorted left to right from most often occurring to least often occurring. TI calculators cannot handle categorical data as such, but using a nominal (number) scale to represent the categories, we can recreate the Pareto Chart which is Figure 2-7 in the text. The chart at right reproduces the data from the text on top complaints to the FCC against U.S. telephone carriers.

Complaint	Frequency
Slamming	12478
Rates	4473
Cramming	1214
Marketing	1007
Intl. Calling	766
Access Charges	614
Operators	534

Since there are seven categories of complaint, enter the numbers 1 through 7 in L1, and the number of complaints in L2. It is important that these be entered in descending order for the Pareto chart.

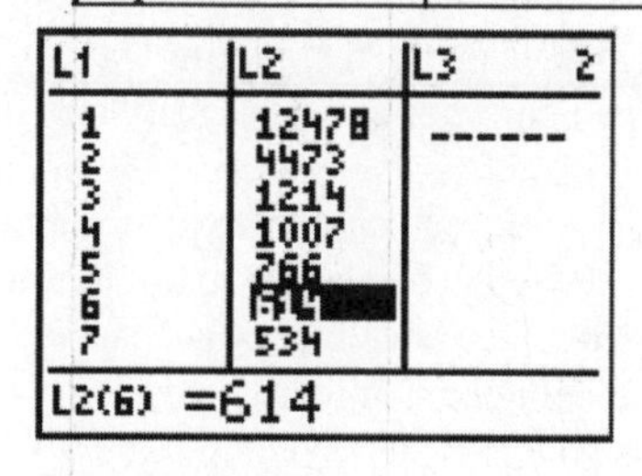

Set up the plot as a histogram with tabled data (use L1 as the Xlist and L2 as the frequency list). We want to see all the bars, and also specify that each bar has width 1. Ymin has been set to a low number so that using [TRACE] does not obscure the final plot. The same window settings will work on the TI-89, but be sure to set the histogram bucket size to 1.

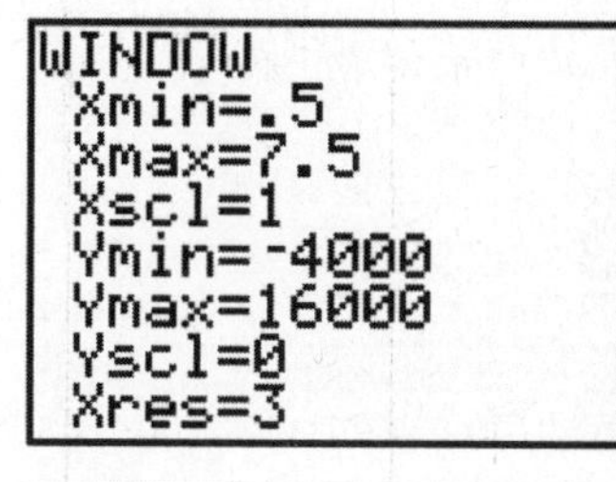

Pressing [GRAPH] followed by [TRACE] shows the following. Visually one can see slamming is clearly the most frequent complaint. We also see the relative equality of complaint types 4 and 5 (Cramming and Marketing) as well as relative equality of the last two complaint types.

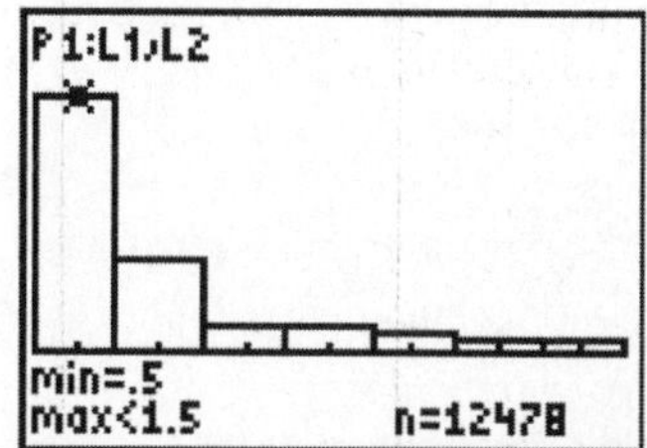

PIE CHARTS

Pie charts are not among the statistical plot types for the TI-83 Plus. However, the calculator can aid in the making of these plots by calculating the degree measure of the center angle (portion of the entire 360° circle) for the pie wedge representing each category. We illustrate this using the data set of phone company complaints. The instructions are analogous for both the 83/83 and 89 calculators.

With the data on frequency of each complaint type already in L2, highlight list name L3. Type 360L2[÷]sum(L2 as in the bottom of the same screen with "sum(" being pasted in by pressing [2nd] [STAT][▶] [▶] [5] (It is the fifth choice on the LIST<Math> menu). On the TI-89, access the List menu with [F3].

In the screen at right, we see that slamming accounts for more than half of all complaints to the FCC. The central angle for its slice of the pie would be 213 °.

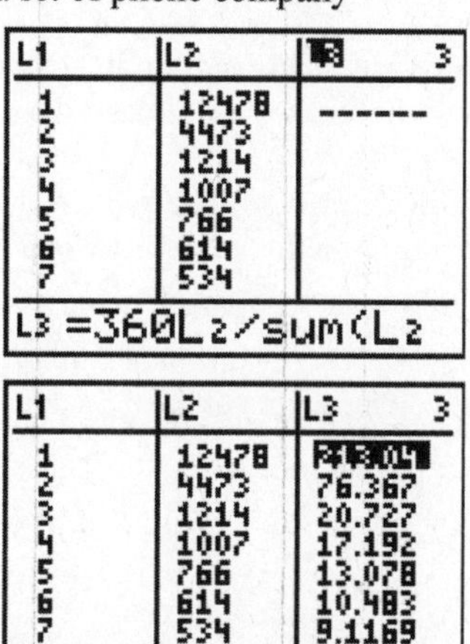

SCATTERPLOTS

Scatterplots will be discussed in Chapter 10 on Linear Regression.

TIME-SERIES GRAPHS

Year	Screens
1987	2050
1988	1500
1989	1000
1990	900
1991	875
1992	860
1993	800
1994	850
1995	825
1996	800
1997	750
1998	710
1999	700
2000	600

Time-series graphs show how a single variable has changed with the passage of time, as opposed to histograms or frequency polygons which show a "snapshot" of the distribution of the variable at one specific time. These are constructed as connected plots, just as frequency polygons are. The time index is always displayed on the *X*-axis, and the variable of interest on the *Y*-axis. The table at right should help us recreate the graph on the demise of drive-ins shown in the text.

To enter the years in sequence, one could type them all in, or enter the last two digits (calling 2000 year 100), but the easiest way to enter sequential data uses the seq(function which is choice 5 on the List Ops menu. The command works the same way on all TI calculators.

```
L1      L2      L3    2
1987    2050    ------
1988    1500
1989    1000
1990    900
1991    875
1992    860
1993    800
L2(1)=2050
```

Move the cursor so that the name of list L1 (list1) is highlighted. Go to the List menu ([2nd][STAT] on the 83/84 or [F3] on the 89), arrow to Ops, then select menu choice 5. Type the parameters for the command: [X][,][X][,][1][9][8][7][,][2][0][0][0][)]. Pressing [ENTER] populates the list. Now enter the number of drive-in screens in L2. The screen at right shows the results.

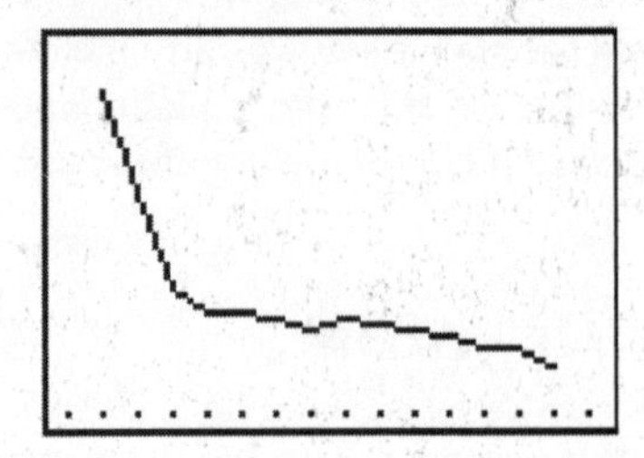

My statistics plot was defined just as we did for frequency polygons: a connected scatterplot with the Xlist being L1, the Ylist L2. I also selected to use the single pixel point mark so as not to obscure the graph. Pressing [F5] on the 89 or [ZOOM][9] on an 83/84 displays the plot at right.

WHAT CAN GO WRONG?

There are several common error messages which might show and "foul-ups" which can occur when graphing data. In this section, I describe those most frequently encountered.

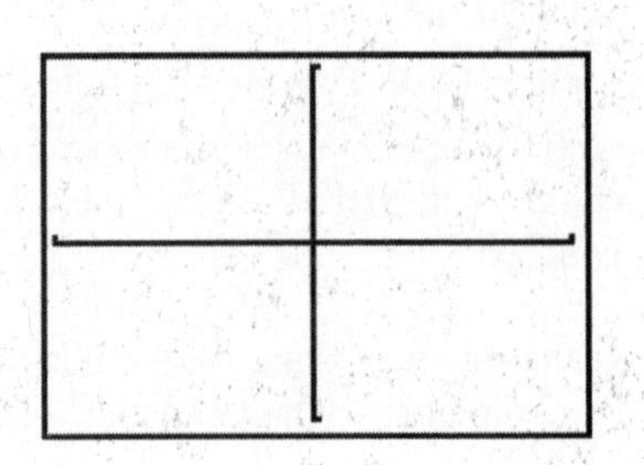

Help! I can't see the picture!
Seeing something like this (or a blank screen) is an indication of a windowing problem. This is usually caused by pressing [GRAPH] using an old setting. Try pressing [ZOOM][9] to display the graph with the current data. This error can also be due to having failed to turn the plot "On."

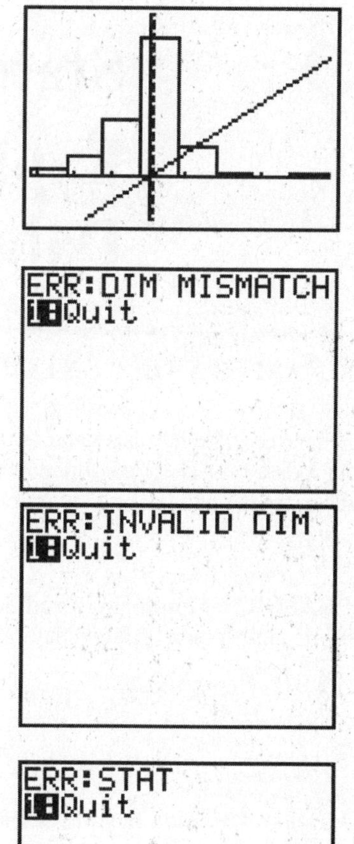

What's that weird line (or curve)?

There was a function entered on the [Y=] screen. The calculator graphs everything it possibly can at once. To eliminate the line, press [Y=]. For each function on the screen, move the cursor to the function and press [CLEAR] to erase it. Then redraw the desired graph by pressing [GRAPH].

What's a Dim Mismatch?

This common error results from having two lists of unequal length. Here, it pertains either to a histogram with frequencies specified or a time plot. Press [ENTER] to clear the message, then return to the statistics editor and fix the problem.

What's an Invalid Dim?

This problem is generally caused by reference to an empty list. Check the statistics editor for the lists you intended to use, then go back to the plot definition screen and correct them. On the TI-89 this error message is Dimension.

Stat?

This error is caused by having two stat plots turned on at the same time. What happened is the calculator tried to graph both, but the scalings are incompatible. Go to the STAT PLOT menu and turn off any undesired plot.

Plot setup?

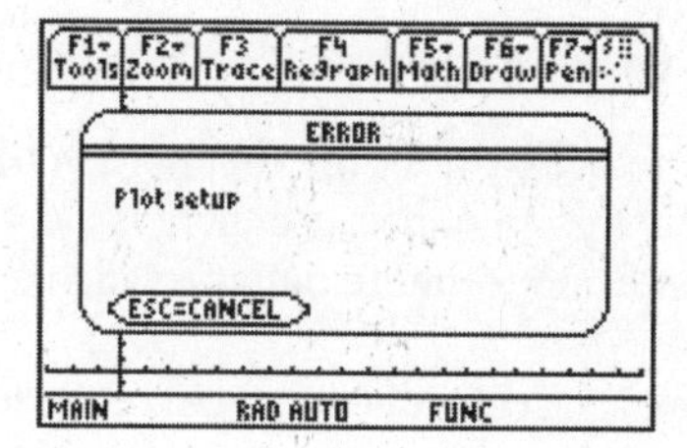

This TI-89 error is caused by having two stat plots turned on at the same time. What happened is the calculator tried to graph both, but the scalings are incompatible. Go to the Stat plots menu and turn off any undesired plot by moving the cursor to that plot, and pressing [F4].

3 Statistics for Describing, Exploring, and Comparing Data

MEASURES OF CENTER

A measure of center is a value at the center or middle of a data set. They describe a "typical" middle value for a data set. There are four common measures of center: the mean, median, mode, and midrange. All of these are at one time or another referred to as the "average." You should be specific which measure you are using.

EXAMPLE Monitoring Lead in Air: Listed below are 6 measurements of the lead in the air (measured in $\mu g/m^3$) taken on 6 different days after Sept. 11, 2001 at Building 5 World Trade Center.

5.40 1.10 0.42 0.73 0.48 1.10

Mean and Median from Raw Data – TI-83/84

Enter the data in L1. Press STAT, then press ▶ to move to the CALC submenu, then press 1 for `1:1-Var Stats`, then press 2nd 1 to name the list L1 as the list for which you want the statistics.

```
1-Var Stats L1
```

Press ENTER for the first portion of the output. You know this is only the first portion, because there is the ↓ on the lower left corner, which is an indicator of more output available. From this first screen, we find the mean (arithmetic average) of the lead readings is $\bar{x} = \frac{\sum x}{n} = 1.538$ (and change, but we always report one more decimal place than was in the original data, so three here).

```
1-Var Stats
 x̄=1.538333333
 Σx=9.23
 Σx²=32.5197
 Sx=1.914203925
 σx=1.747421116
↓n=6
```

To reveal the second output screen, hold down the ▼ key. Here we find the median = Med = .915. Many of the other statistics displayed on the two screens will be discussed later.

```
1-Var Stats
↑n=6
 minX=.42
 Q1=.48
 Med=.915
 Q3=1.1
 maxX=5.4
```

Mean and Median from Raw Data – TI-89

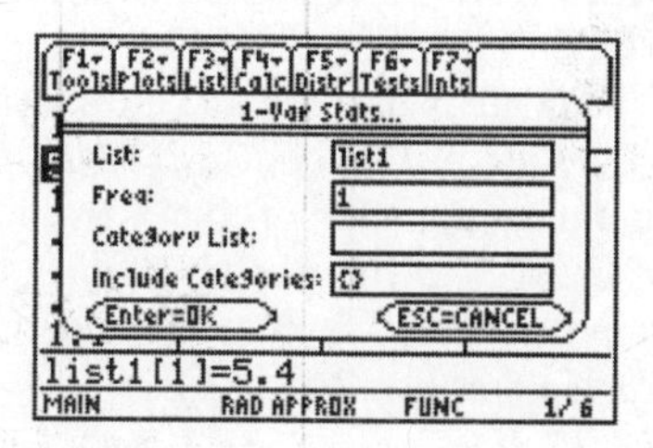

Enter the data in `List1`. Press [F4] (`Calc`), then press [ENTER] to select `1:1-Var Stats`. Press [2nd] [-] (`VAR-LINK`) to and find the list name `List1` as the list for which you want the statistics. Press [ENTER] to paste the list name into the input screen. `Freq` should be set to 1, as each value in the list occurred once. Your input screen should look like the one at right.

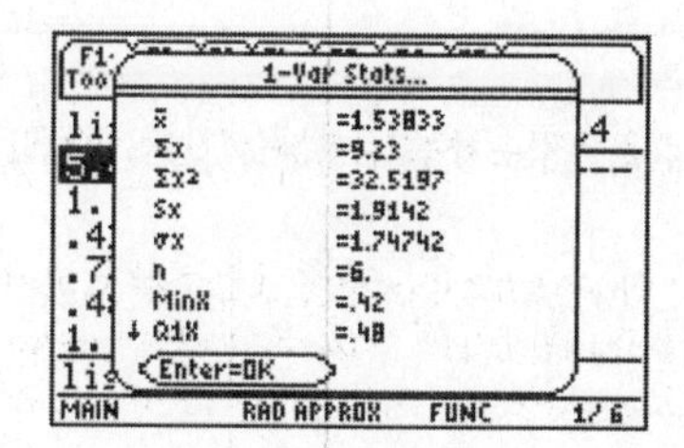

Press [ENTER] for the first portion of the output. You know this is only the first portion, because there is the ↓ on the lower left corner, which is an indicator of more output available. From this first screen, we find the mean (arithmetic average) of the lead readings is $\bar{x} = \frac{\sum x}{n} = 1.538$ (and change, but we always report one more decimal place than was in the original data, so three here).

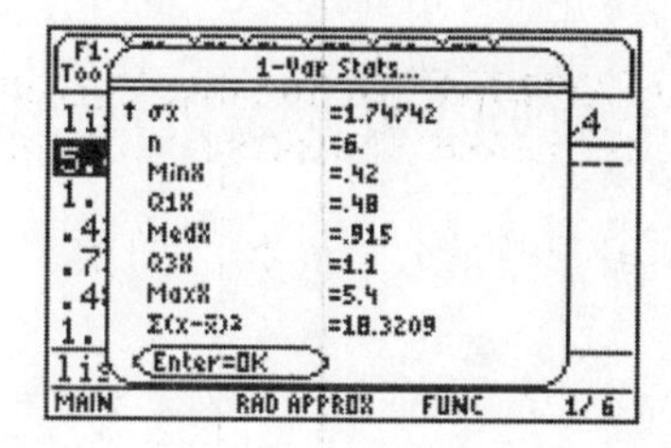

To reveal the second output screen), hold down the [▼] key. Here we find the median = Med = .915. Many of the other statistics displayed on the two screens will be discussed later.

Mode

TI calculators do not automatically calculate the mode of a data set. Ordering the data and counting the frequency of each value, as discussed in the section on dotplots, will aid in finding the mode (if one exists). The modal class can be found by looking for the class with the highest frequency in a frequency table. (See page 22 of this manual.)

Midrange

EXAMPLE Monitoring Lead in Air: Find the midrange of the six data values

5.40 1.10 0.42 0.73 0.48 1.10

Two of the values on the second output screen are `Min` and `Max`. These are the lowest and highest values in the data set. We can use the output displayed above to calculate midrange $= \frac{\text{Max} + \text{Min}}{2} = \frac{5.4 + 0.42}{2} = 2.91$.

Mean from a Frequency Distribution

Age	Frequency	Midpoint
21-30	28	**25.5**
31-40	30	**35.5**
41-50	12	**45.5**
51-60	2	**55.5**
61-70	2	**65.5**
71-80	**2**	**75.5**

When data is presented as a frequency distribution, the actual data values are lost, but one can approximate the mean and other summary statistics. Table 3-2 which gives the frequency distribution of the ages of Best Actress winners is reproduced here for convenience. Enter the age midpoints in `L1` (`list1`) and the frequencies in `L2` (`list2`).

On a TI-83/84, press STAT then ▶ to CALC and press ENTER to select `1:1-Var Stats`. We need to tell the calculator to use both lists, but we still want statistics for only one variable (the actresses' ages). To do this, specify the list names `L1`, `L2` (2nd 1 , 2nd 2) as at right.

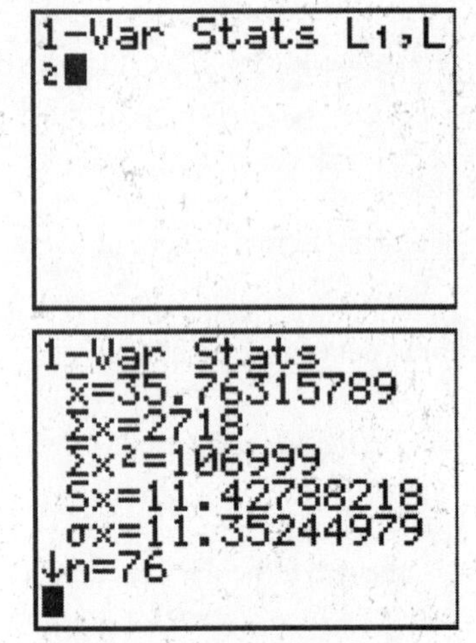

Press ENTER to see the results. . We see that the mean value calculated from the frequency table is 35.8 (rounded to one more place than was in the original data) which is slightly larger than the true mean of the data, 35.7.

On a TI-89 calculator, to use frequency tables in the calculation, use `1-Var Stats` from the `Calc` menu as before, but specify the second frequency list in the box labeled `Freq`. See the example at right.

Weighted Mean

EXAMPLE: Find the mean of three tests scores (85, 90, 75) if the first tests counts 20%, the second counts 30%, and the third test counts for 50% of the final grade.

Put the scores in `L1` (`list1`) and the weights in `L2` (`list2`), and compute as if this were a frequency distribution as described above. This student has a grade of 81.5, so a B (at least in my grading system).

MEASURES OF VARIATION

A measure of variation indicates in some manner how spread out or close together a set of data might be. Data sets with large variability indicate less consistency in the values.

EXAMPLE : The waiting times (in minutes) of three Mulberry Bank customers are 1, 3, and 14. Find the range, standard deviation and variance of this data.

Standard Deviation

With the data in L1, press [STAT], arrow to CALC, press [ENTER] to select 1:1-VarStats and enter the list name ([2nd][1]). The standard deviation is Sx = 7 minutes. The σx value of 5.7 would be used if we had data for the entire population (all waiting times at the bank, but we don't have this - just a sample.) On a TI-89, use 1-Var Stats as described above.

```
1-Var Stats
x̄=6
Σx=18
Σx²=206
Sx=7
σx=5.715476066
↓n=3
```

Variance

The variance is not among the statistics given by the 1-Var Stats display, but it is easy to calculate since it is the square of the standard deviation. In this case, Variance = $(Sx)^2 = 7^2 = 49$ min^2.

Range

The range is given by maxX - minX = 14 – 1 = 13 minutes.

Using Statistics Variables

Sometimes it is useful to be able to recall statistics *that have already been calculated.* One example of this is in computing the variance of a distribution. If there are many digits in the standard deviation, retyping them could lead to typos and erroneous results. We also want to keep as many decimal places as possible in intermediate calculations and do any rounding at the final result. It is important to know that you can only paste a statistics variable after you have performed 1-Var Stats on your *current* data set, otherwise the statistics stored will be from some past calculation.

TI-83/84 Procedure

Press [VARS] for the VARS menu screen, then press [5] for the Statistics sub menu screen at right.

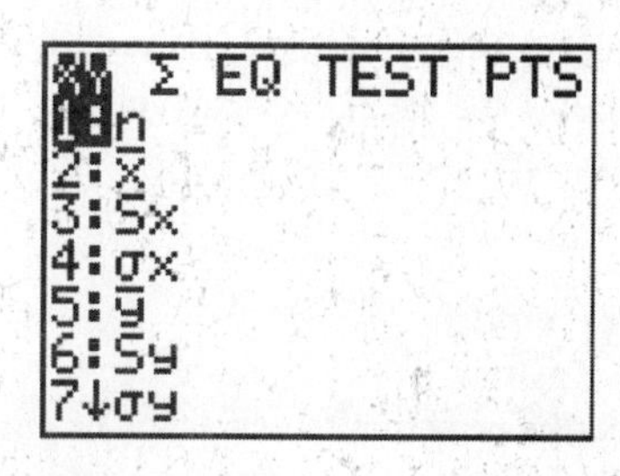

Press [3] and Sx is pasted onto your home screen. Now, press [ENTER] to see that Sx is 7 as in the top of my screen. Press [x²] and then [ENTER] to again see variance = 49.

```
Sx
                 7
Ans²
                49
```

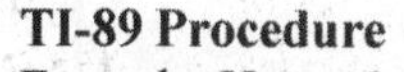

TI-89 Procedure

From the Home screen, press [2nd][–] ([VAR-LINK]). The MAIN folder is normally expanded. Arrow down to highlight the folder name, and press ◀ to collapse that list. You should see a screen like the one at right.

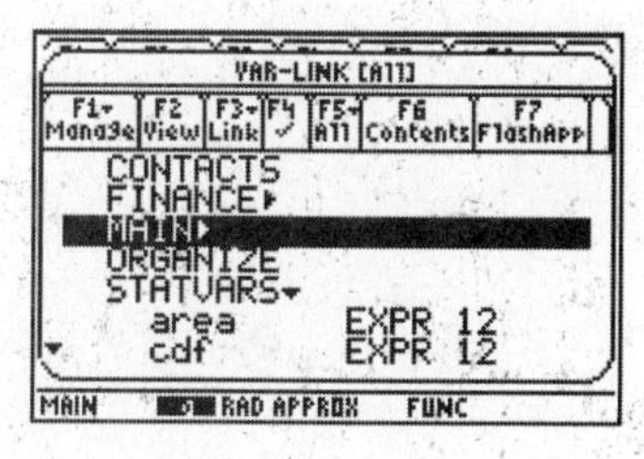

Arrow down to find the statistics variable you want. The sample standard deviation is `sx_`. Arrow down to find and highlight it, then press [ENTER] to transfer the variable name to the home screen. Press [ENTER] again to display the value. Now (to find the variance) press [^][2] to square the standard deviation and compute the variance. My completed calculation is displayed at right.

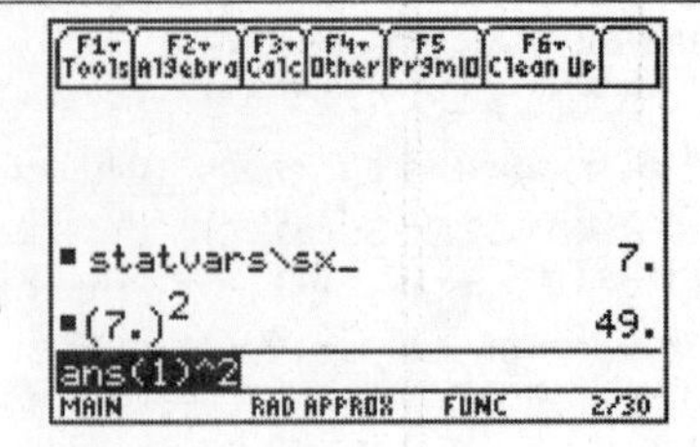

Standard Deviation from a Frequency Distribution

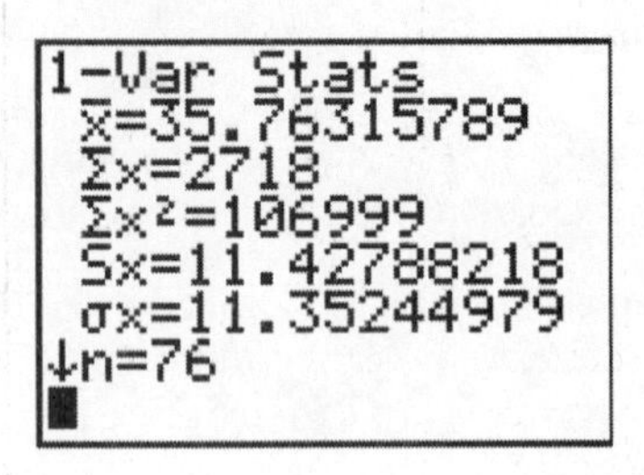

On page 27 we estimated the mean age of the Best Actress winners using a frequency distribution. Our output screen is duplicated at right. From this we estimate the standard deviation of the ages is 11.4. Bear in mind that just as the mean in this instance is an estimate of the true value that would be obtained from the entire data set, so is the standard deviation an estimate. (The standard deviation using all the data is 11.1.)

MEASURES OF RELATIVE STANDING

Quartiles and Percentiles

Quartiles are displayed on the `1-Var Stats` output screen after pressing [▼] several times to scroll to the bottom. Your text illustrates the different implementations for finding quartiles using data values 1, 3, 6, 10, 15, 21, 28, and 36. I have entered those values in `list1` and computed the summary statistics. From this screen we find the first quartile (`Q1X`) is 4.5 and the third quartile (`Q3X`) is 24.5. TI calculators all use the algorithm of finding the median of each half of the data set after it is split by the median.

Percentiles

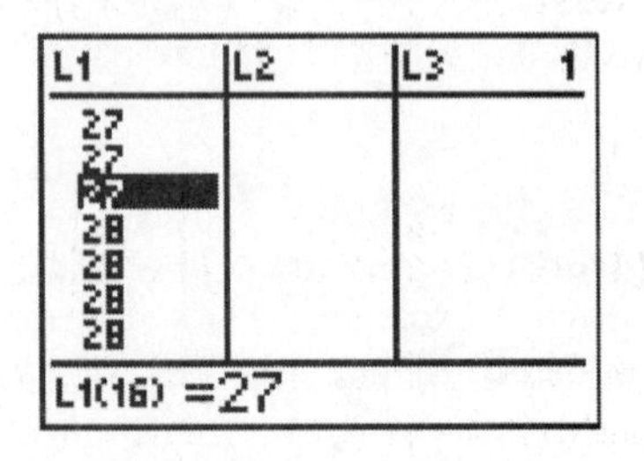

Find the percentile corresponding to a Best Actress winner who was 30 years old.
We must first sort the data in ascending order (see Chapter 1). Now scroll down the list to find the first occurrence of 30. The first 30-year-old is the 27th entry, so there are 26 winners who were younger. There are 76 total winners in our list, so we calculate (26/76)*100 = 34.2. Thus, an age of 30 is the 34th percentile of this distribution.

Find the value of the 20th percentile or P_{20}.
We calculate (20/100)*76 = 15.2. Since this is not a whole number, we round up to the nearest whole number which is 16. Thus the 20th percentile is the 16th value in our sorted list. We find this value to be an age of 27.

Find the third quartile Q_3.
If we look back at the calculator's Q_1 statistic displayed in the `1-Var Stats` calculations, we see that the value is `Q3` is 39.5. Alternatively, since $Q_3 = P_{75}$, we could proceed as in the previous example. We find (75/100)*76 = 57. Since 57 is a whole number, our algorithm would then require us to find the value midway between the 57th and 58th data values. After paging through the sorted list, we find these values are 39 and 40. Thus Q3 = (39+40)/2 = 39.5. In this case, our method has yielded the same result as that of the calculator.

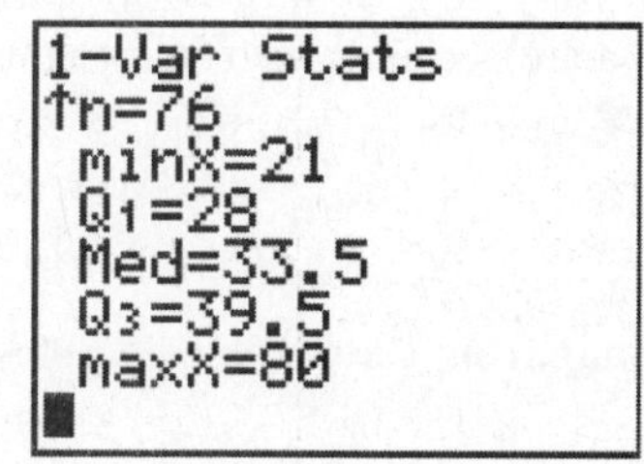

BOXPLOTS AND THE FIVE-NUMBER SUMMARY

Boxplots are another way to picture a distribution. There are two boxplots implemented on TI calculators: the skeletal or regular plot which uses only the five-number summary (min, Q1, med, Q3, and max) to graph the "quarters" of the distribution and the modified boxplot which identifies outliers. They are the fourth and fifth plot types on a TI-83 or -84 (icons for the regular boxplot and for the modified boxplot). On a TI-89 these are choices `3:Box Plot` and `5:Mod Box Plot` on the Plot definition screen. They are usually very good at finding "typos" in a list of data, and they are extremely useful in comparing two distributions.

EXAMPLE Actress' Ages with a typo
I have used the data set of the Best Actress winners' ages, but replaced one occurrence of a 22 with 2222 (a distracted typo). We notice first of all the impact of this typo – the average age of the winner is not 64.6, and the standard deviation is 251! Obviously, thinking about actresses, these are not reasonable values – go back and double check your data entry!

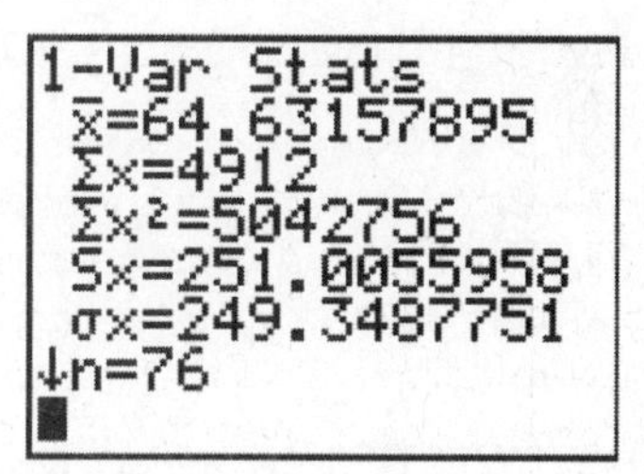

Boxplot (or Box and Whisker Diagram)

I have corrected the typo for this example.

Press [2nd] [Y=] ([F2] on an 89) to get the `Stat Plots` menu and then [1] to get the set-up for `Plot1`. Set up `Plot1` as shown in my screen. We choose the fifth plot type. This is the *skeletal* boxplot which is described in your text.

Press [ZOOM] [9] for the boxplot. Press [TRACE] then use the [◄] and [►] keys to display the **five-number summary**.

Identifying Outliers – The Modified Box Plot

The plot obtained above has a long right-hand tail. It may be that there are outliers present (older winners who don't fit the normal pattern of variation.) TI calculators can create the modified box plot which labels any observation more than 1.5*IQR away from the quartiles as an outlier. In the screen at right, I have changed the plot definition to make a modified box plot.

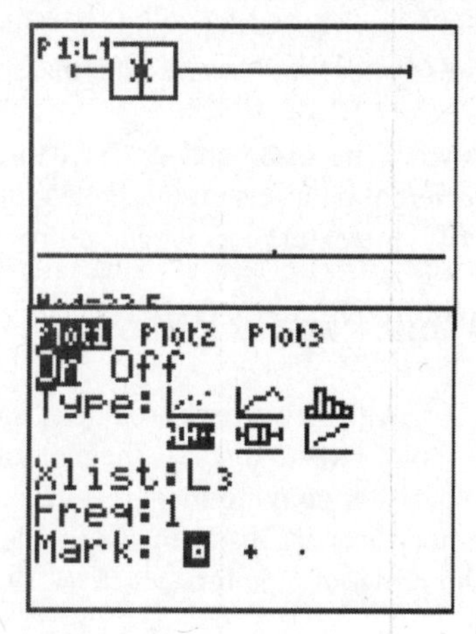

Pressing [ZOOM][9] ([F5] - ZoomData on an 89) and then [TRACE] shows that all actresses older than 54 are unusually old, compared to other winners of Best Actress.

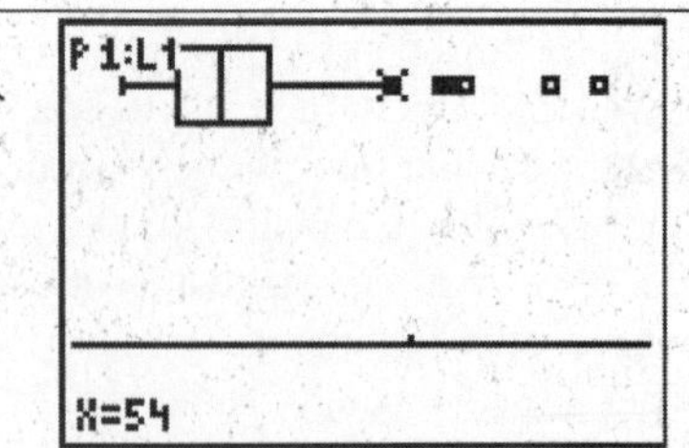

Comparing Data Sets with Side-by-Side Boxplots

We have said before that normally there should be only one statistics plot "on" at a time. Boxplots for comparing two distributions are a good example of a reason to have more than one plot "on" at once.

EXAMPLE Pulse Rates of Men and Women: We compare the pulse rates of 40 females and 40 males as seen in Data Set 1 in Appendix B. We have stored the females' rates in L1 and the males' rates in L2.

Press [2nd] [Y=] to get the Stat Plots menu and then define Plot1 as at right.

Press [2nd] [Y=] to get the Stat Plots menu and then define Plot2 as at right.

Press [ZOOM] [9] to see the two boxplots on a single screen.

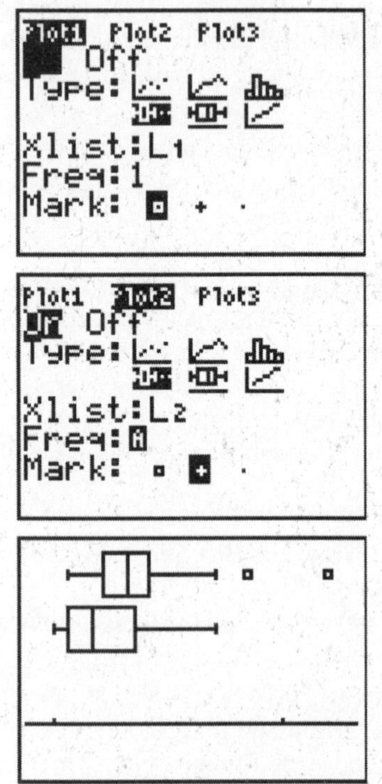

Press [TRACE] then use the [◄] and [►] keys to display the five-number summary. You can use the [▲] and [▼] keys to toggle between the upper plot which is that of the females and the lower plot which is for the males. These side-by-side plots clearly indicate that males, in general, have pulse rates than females and that the pulse rates of females tend to be more variable than those of males.

Note: The TI-83 and -84 have three StatPlots. They can be used to plot box and whisker plots for three different data sets on the same graph. The Ti-89 can display as many as six boxplots at once. Be sure to turn "off" the extra plots when you're done with them!

What Can Go Wrong?

Primarily, errors here are due to data entry mistakes. Always double check what you have entered. If a boxplot fails to display, the plot more than likely was not turned on. Always be sure to turn off any extra plots after copying them to paper. If not, you probably will receive either the Invalid Dim (referring an empty list) or Stat (incompatible window ranges for two plots) error messages. These were discussed in the previous chapter.

4 Probability

In this chapter you will learn how your TI calculator can be used in the calculation of probabilities. There are several built-in functions which make this work much simpler, including the *change to fraction function,* the *raise to power key,* and the *Math Probability menu* which includes options to make the calculation of factorials, combinations and permutations very easy. You will also be introduced to the concept of simulation. This process can be used to approximate probabilities. Your exposure to the process may have the added benefit of giving you a deeper understanding of the concepts of probability.

TWO USEFUL FUNCTIONS

Change to Fraction Function

This function is useful when one wants to display an answer in fraction, rather than decimal form. However, when a fraction is not obvious (say 432/7482), it is better to report the answer in decimal form, using three significant (non-zero) digits.

EXAMPLE Birth Genders: In reality, more boys are born than girls. In one typical group, there are 205 newborn babies, 105 of whom are boys. If one baby is randomly selected from the group, what is the probability that the baby is not a boy?

```
100/205
        .487804878
Ans▸Frac
             20/41
```

Using the Law of Complements, we decide the answer is 100/205=0.488. See the first line of the screen at right. Now, press [MATH] [1] (This chooses the *change to fraction function.)*

Press [ENTER]. You should see the rest of the screen. The result is the

probability written as a reduced fraction $\frac{20}{41}$.

```
F1▾   F2▾     F3▾  F4▾   F5     F6▾
Tools Algebra Calc Other PrgmIO Clean Up

■ 100/205                  .487805
■ exact(100/205)             20/41
exact(100/205)
MAIN     RAD APPROX   FUNC    2/30
```

On a TI-89, we recommended in Chapter 1 that the mode for answers be set to `Approximate` rather than the calculator's default of `Exact`. To obtain exact answers with the calculator in `Approximate` mode, Press [2nd][5] (Math), then press ▶ to expand the `Number` menu. Press [ENTER] to select `exact(`. Now type in the division, being sure to close parentheses before pressing [ENTER] to complete the calculation. The TI-89 equivalent of the TI-83 calculation above is at right.

The Raise-to-the-Power Key
The Probability of "at Least One".

EXAMPLE Gender of Children: Find the probability of a couple having at least one girl among three children. We will assume for the purpose of this example that boys and girls are equally likely. We make use of the complements rule, since all boys is the opposite (complement) of at least one girl.

Find the probability of the complement. P(boy and boy and boy) = $0.5*0.5*0.5 = 0.5^3$.
Type 0.5 [^] [3]. Then press [ENTER] for 0.125.

Note: Keep in mind the answer you have now is the complement of the answer you seek.

Press [1] [−] [2nd] [(-)]. This takes your answer from above and subtracts it from 1. Press [ENTER] for 0.875. This could be changed to the fraction 7/8 as shown.

```
.5^3
                .125
1-Ans
                .875
Ans▸Frac
                 7/8
```

PROBABILITIES THROUGH SIMULATION

Finding probabilities of events can sometimes be difficult. We can often gain knowledge and insight into the problem by developing a simulation of it. The techniques used in the examples in this section build upon one another. Thus each example assumes you are familiar with the ones preceding it.

EXAMPLE Gender Selection: When testing techniques of gender selection, medical researchers need to know probability values of different outcomes, such as the probability of getting at least 60 girls among 100 children. Assuming that male and female births are equally likely, describe a simulation that results in the gender of 100 newborn babies.
We will perform this simulation by generating 100 random 0's and 1's. Each will be equally likely.

TI-83/84 Procedure:

We begin by setting the "seed" for our random number generator as 136. (The reason for this step is to set your calculator, so it will generate the same random numbers as generated by the calculator used in this manual. This step is **not** necessary when performing your own simulations.) Press 136 [STO▸] [MATH], arrow to PRB, press [1] to select `rand`, then [ENTER] for the top of my screen.

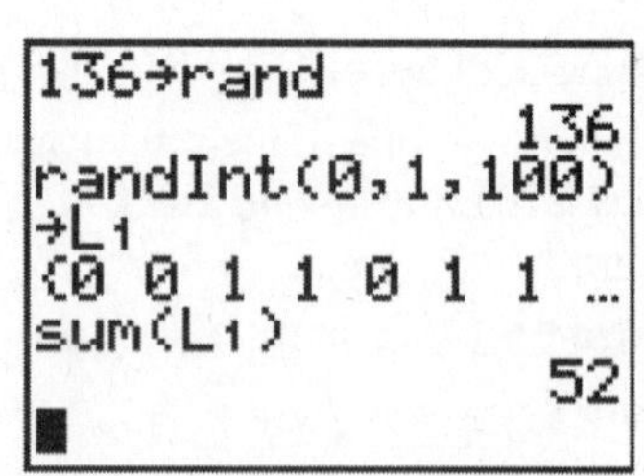

Type [MATH] [◄] [5] to get the `randInt` prompt on the main screen. Then type 0,1,100 [STO▸] [2nd] [1] [ENTER] to have the TI-83/84 generate one hundred 0's and 1's and store them in `L1`.
We can see the results which begin with {0 0 1 1 0 1 1, …} These can be translated as {B B G G B G G…}if we let the 1's represent girls and the 0's represent boys.

In the last line of the screen above, we summed the elements in list `L1` to find out how many 1's (girls) we had in this simulation of 100 births. Press [2nd] [STAT] for the `LIST` menu, then [►] to `Math`, select menu option `5:sum(`, then [2nd] [1] [ENTER] to sum the elements of list `L1`. We see that this time we had 52 girls.
Note: We did not have 60 or more girls in this simulation. In order to get an idea of the probability of at least 60 girls in 100 births, you would have to perform the simulation repeatedly, keeping track of how often the event of 60 or more girls occurred.

TI-89 Procedure:

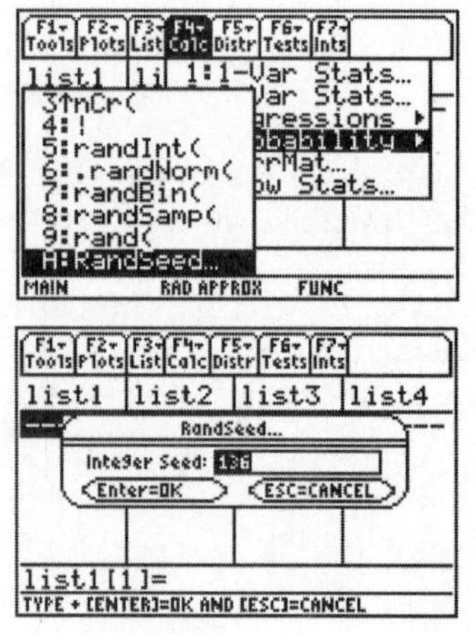

With the calculator in the Statistics/List Editor application, press [F4] (`Calculate`) and either arrow to `4:Probability` or press [4]. To set the seed for the random number generator, the menu option is `A:RandSeed`. You can arrow down to find it, but it is easier to press ⊖ once.

Type in the desired seed, and press [ENTER].

Now, move the cursor to highlight the name of an empty list, press [F4] (Calculate). Select menu option 4:Probability as before, but now select option `5:randInt(`. Now type 0,1,100) - remember to close the parentheses - and press [ENTER] to populate the list.

To sum the list, move the cursor to the top of an empty list. Press [F3] (`List`), arrow to menu option 3:Math and press [▶] to expand the menu. Select menu option `5:sum(`. Now we need to tell the calculator which list to sum. Press [F3] again, then [ENTER] to select `1:Names`. Find the name of the appropriate list, move the highlight to it, close the parentheses, and then press [ENTER] to complete the calculation.

EXAMPLE Same Birthday: One classic exercise in probability is the *birthday problem* in which we find the probability that in a class of 25 students at least 2 students have the same birthday. Ignoring leap years, describe a simulation of the experiment that yields birthdays of 25 students in a class.

We will perform this simulation by generating 25 random integers between 1 and 365 (representing the 365 possible birthdays in a non-leap year.)

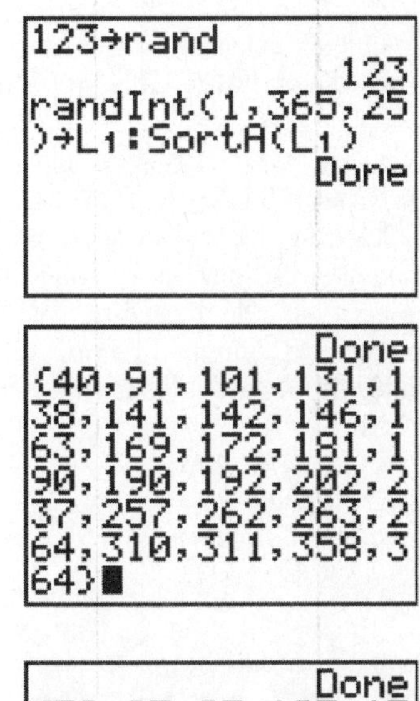

Again we set the seed so as to obtain the same outputs as this manual. Press 123 [STO▶] [MATH] [◀] [1] [ENTER].

Next generate 25 random integers between 1 and 365, store them in `L1` and sort `L1`. Do this by typing [MATH] [◀] [5] to get the `randInt(` prompt then typing 1, 365, 25 [STO▶] [2nd] [1] [ALPHA] [.] [STAT] [2] [2nd] [1] [ENTER] to see the `Done` message as in my screen. Note the [ALPHA] [.] sequence yielded a colon on the screen. The colon can be used to tie several statements together in one command.

We wish to see the results of our simulation, so recall `L1` to the screen with [2nd] [STO▶] [2nd] [1] and then press [ENTER] for screen the screen at right. We can see that two students in this simulated class had the same birthday on the 190th day of the year.

To repeat this simulation without retyping all the commands, simply press [2nd] [ENTER] [ENTER]. This sequence yields the ENTRY command which repeats the last command line. Now press [2nd] [STO▶] [2nd] [1] to recall the new list `L1` as in the screen at right. Press [ENTER] to check the next set of results. It shows no birthday matches in this simulated group of 25 students.

To do this on a TI-89, follow the steps outlined above using the 89 procedure previously described. To sort the list, place the cursor so it highlights the list name, press [F3] (`List`), arrow to option `2:Ops`, then press ▶ to expand the menu. Press [ENTER] to select `1:Sort List`. The dialog box should look like the one at right (assuming your random numbers were in `list1`.) Page through the list looking for any duplicates.

```
F1- F2- F3- F4- F5- F6- F7-
Tools Plots List Calc Distr Tests Ints
list1 list2 list3 list4
40.        Sort List...
91   List:        list1
10   Sort Order:  Ascending→
13   (Enter=OK)   (ESC=CANCEL)
138.
141.
list1[1]=40.
MAIN     RAD APPROX     FUNC
```

Thus far we have seen a 50% chance of a birthday match in a group of 25. Naturally, the simulation should be repeated many more times to get a better estimate of the actual probability of this event.

EXAMPLE Simulating Dice: Describe a procedure for simulating the rolling of a pair of six-sided dice.

This time we seed the random number generator with 4321. (Again, this step is only necessary if you are trying to duplicate the results presented in this guide.)

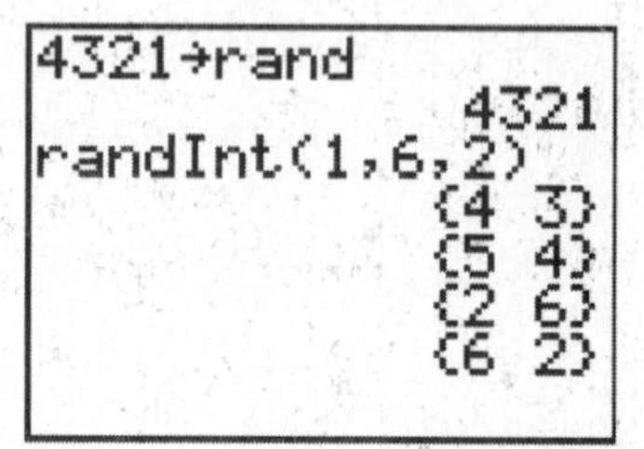

To generate two random integers between 1 and 6 type [MATH] [◄] [5] to get the `randInt(` prompt then type 1, 6,2 and close the parentheses. Press [ENTER]. If you used the seed from step 1, you should see the result (4,3) as at right. Continue to press [ENTER] to generate new rolls as shown in the bottom of the screen. TI-89 calculators cannot perform this type of random integer generation on the home screen, but must use statistics lists with the `randInt(` function. See the examples in Chapter 1.

To quickly generate 100 sets of two rolls and calculate their sums, you could proceed as follows. This procedure is analogous for all calculators.

Generate 100 integers between 1 and 6 and store in `L1` (`list1`).
Generate 100 integers between 1 and 6 and store in `L2` (`list2`).
Add `L1` and `L2` and store the results in `L3`. On TI-83/84 calculators, this can either be done on the Home screen as at right, or by highlighting the new list name in the statistics editor and typing [2nd][1][+][2nd][2][ENTER] to execute the command `L1+L2`. On a TI-89, since the generation must be done within the statistics editor, it is easiest to perform the addition in the editor.

```
randInt(1,6,10)→
L1
{5 4 5 3 2 5 4 ...
randInt(1,6,100)
→L2
{5 2 5 3 4 4 5 ...
L1+L2→L3
```

Press [STAT] [1] to see the results in your lists.

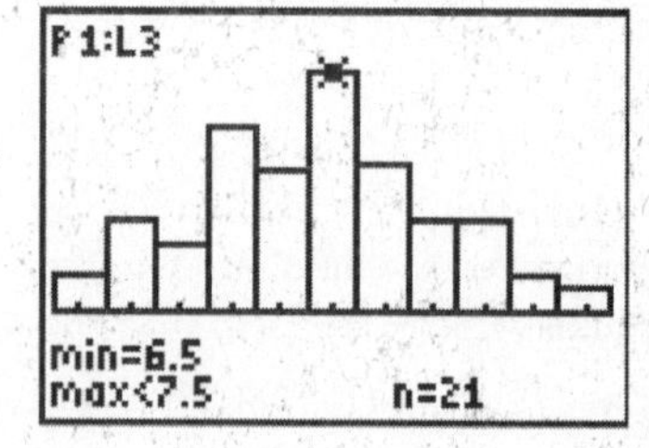

The screen at right shows a histogram of my results from this simulation. This histogram is for the data set consisting of the sums of the two rolls. This set was stored in `L3`. The Window settings were `Xmin` = 1.5, `Xmax` = 12.5, `Xscl` = 1, `Ymin` = -8, `Ymax` = 25. This simulation could be used to approximate the probability of rolling a certain sum. For example, we can see that our 100 rolls yielded 21 sums of 7. Thus we could estimate the probability of rolling a 7 to be 21%. (In actuality the value is 1/6 or 16.7%. More simulations would no doubt lead us closer to the truth.) Your results will vary.

COUNTING

In many probability problems, the big obstacle is trying to determine the number of possible (or favorable) outcomes. The procedures in this section can help.

Factorials

Notation: The factorial symbol is !

EXAMPLE Routes to Rides: How many different ride orders are possible if you want to visit Space Mountain, Tower of Terror, Rock 'n Roller Coaster, Mission Space and Dinosaur at Disney World on your first day?

By applying the factorial rule, we know that the 5 rides can be arranged in a total of 5! ways. We must calculate 5!. Do so on a TI-83 or -84, first type 5. Then press [MATH], arrow to PRB, and select option 4: ! Which is the factorial (!) symbol. Press [ENTER] to see the results of the calculation.

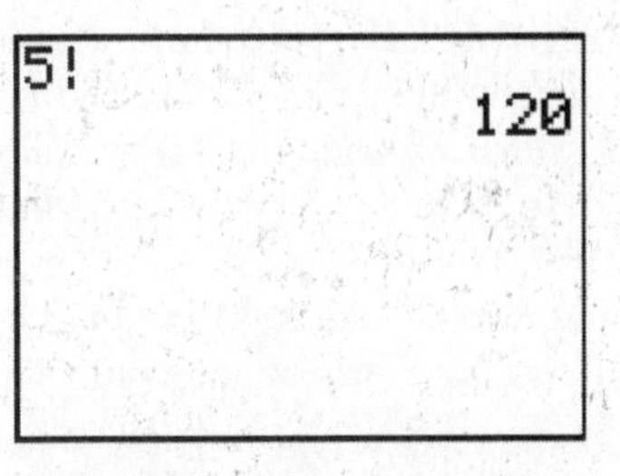

On a TI-89, the procedure is analogous, but keys are not the same, due to menu differences. On the home screen, type the 5 as above. Now, press [2nd][5] (Math), [7] (Probability), then [ENTER] to select option 1: !. From the screen shown, press [ENTER] to transfer the factorial symbol to the input area, then [ENTER] again to complete the calculation.

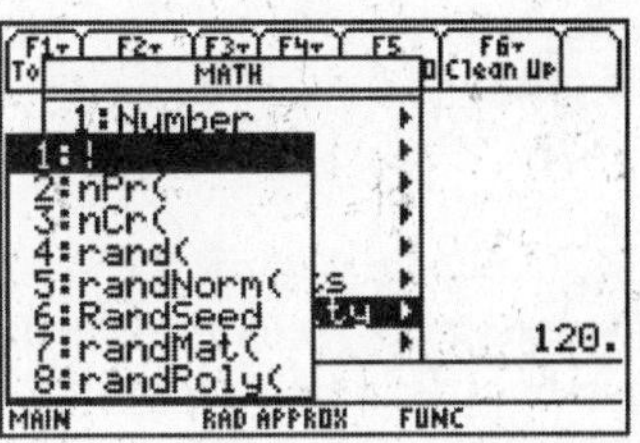

Permutations - All Items Distinct

EXAMPLE State Capitals: You want to conduct surveys in state capitals, but have only time enough to visit four of them. How many possible routes are there?

We know that we need to calculate the number of permutations of 4 objects selected from 50 available objects. Begin by typing 50. Press [MATH] [◄] [2] to select nPr from the Math, Probability submenu. Then type 4.

Press [ENTER] to see the result 5,527,200. I think you need to add a few other conditions in planning your trip!

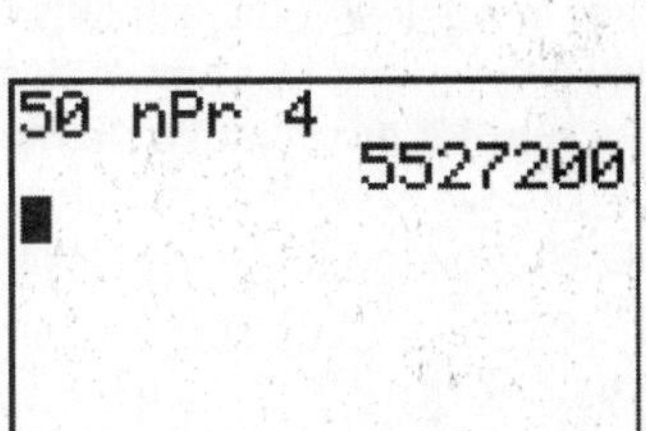

Permutations are also option 2 on the Probability submenu on the TI-89, but one enters the command first, then the parameters n (the total to choose from) and k (the number desired) separated by a comma. Be sure to close the parentheses.

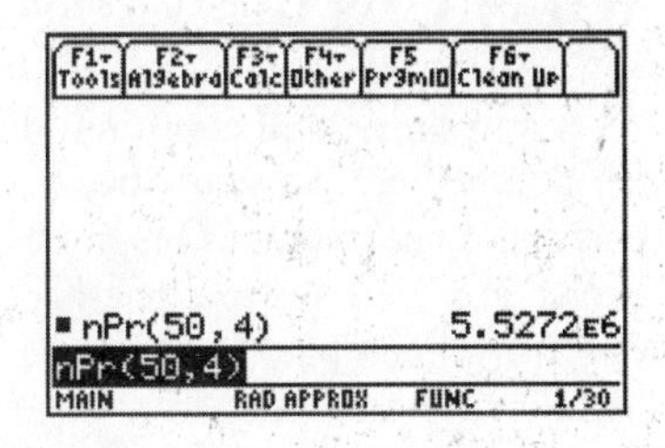

Permutations - All Items not Distinct

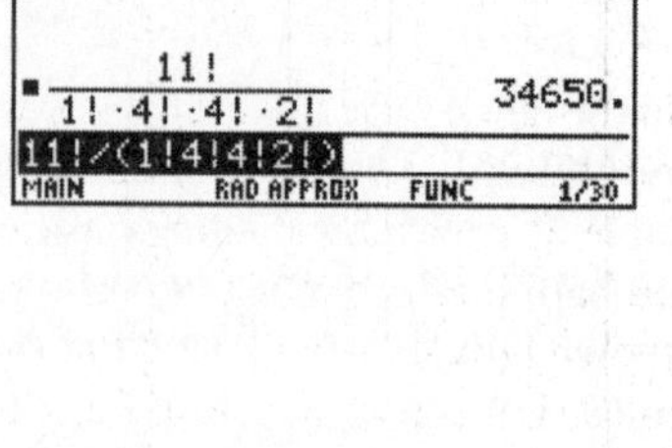

EXAMPLE Mississippi: How many distinct orderings are there of the letters in the word Mississippi? Questions like this are classics in the world of Probability. First, we note that there are 11 total letters in the word: 1-M, 4-i's, 4-s's, and 2-p's. Since one cannot (normally) distinguish one letter i from another, the total number of *distinct* permutations is 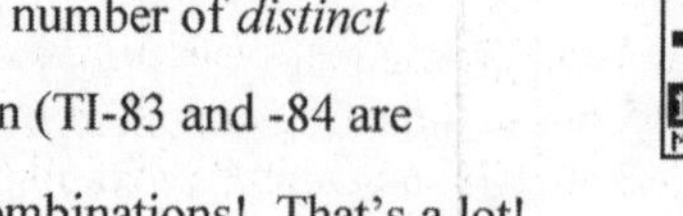 $\frac{11!}{1!4!4!2!}$. My TI-89 calculation (TI-83 and -84 are analogous) shows there are 34,650 distinct combinations! That's a lot!

Combinations

EXAMPLE Phase I of a Clinical Trial: When testing a new drug on humans, a clinical test is normally done in three phases. Phase I is conducted with a relatively small number of healthy volunteers. Let's assume we want to treat 8 healthy humans with a new drug, and we have 10 suitable volunteers. If we want to treat all 8 patients at once, how many different treatment groups are possible?

```
10 nCr 8
                  45
```

We wish to select 8 from the available 10. But we are not concerned with the order of selection, merely the final result (the treatment group). On the home screen, type 10, then press [MATH], arrow to PRB and press [3] to select option 3:nCr. Now type in 8 and press [ENTER]. We see there are 45 potentially distinct groups.

Maine Lottery: In the Maine lottery, a player wins or shares in the jackpot by selecting the correct 6-number combination when 6 different numbers from 1 through 42 are drawn. If a player selects one particular 6-number combination, find the probability of winning the jackpot. (Order is irrelevant).

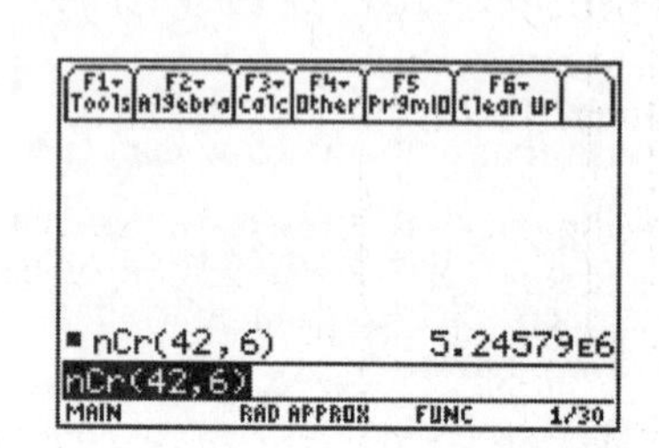

Because 6 different numbers are selected from 42 and order is irrelevant, we know we must calculate the number of combinations of 6 objects chosen from 42. This will tells us the total number of possible lottery outcomes. On a TI-83 or -84 we would do this as in the example above entering 42, then finding nCr and following that with a 6 before pressing [ENTER]. On a TI-89 home screen , we first locate the command, which is option 3 on the Math, Probability submenu. Press [2nd][5] (Math), [7] (Probability), [3] (nCr) to transfer the shell to the input area, then type 42,6) [ENTER] to complete the command. There are 5,245,790 possible winning combinations! No wonder lotteries are called a "tax on people who flunked statistics!"

WHAT CAN GO WRONG?

How can there be 0 combinations?

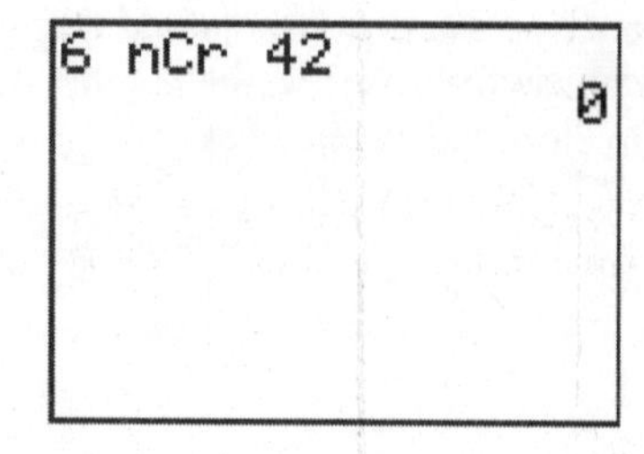

There can, but it's usually because you entered the parameters for the command backwards. One might be thinking of (in this example) choosing 6 from 42, but that's not how it works. The binomial coefficient (number of combinations) is read "n choose k." This helps keep straight that the total to choose from comes first.

5 Discrete Probability Distributions

In this chapter you will learn how your TI calculator can be used when working with probability distribution functions. First, you will learn how to use the calculator when given a probability function in the form of a table of values for the random variable with associated probabilities. You will also learn how to calculate probabilities of random variables which have the binomial or poisson distribution. In both cases, you will be shown how to calculate on the Home screen using the formulas for these distributions and alternatively how to use built-in distribution functions in the DISTR menu. Either alternative alleviates the need to use probability tables and, in fact, yields answers with more accuracy than the tables.

PROBABILITY DISTRIBUTIONS BY TABLES

EXAMPLE Jury Pools: Table 4-1, reproduced below, describes the probability distribution for the number of Mexican-Americans on a jury of 12, assuming that jurors are randomly selected form a population in which 80% of the eligible people are Mexican-American.

X	0	1	2	3	4	5	6	7	8	9	10	11	12
P(x)	0.0+	0.0+	0.0+	0.0+	0.001	0.003	0.016	0.053	0.133	0.236	0.283	0.206	0.069

Probability Histogram

To plot the probability histogram from the table follow these steps:

Put the X values in L1 and the P(x) values in L2. (Type 0, not 0+ for the first four probability values.)
Note: Sum(L2) = 1 as expected.

Set up Plot1 as at right. Notice that since each value of the variable does not represent one observation, we have changed Freq: from 1 to L2 which contains the probabilities.

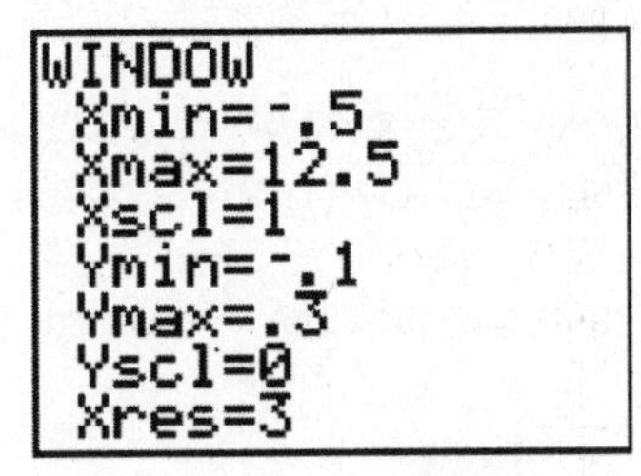

Since we clearly want to see a bar for each potential value of the variable X (the number of Mexican-American jurors), set the WINDOW values for this histogram as at right. Each bar has width 1, and Ymax is set to .3 (slightly larger than the largest probability) and Ymin is set to -.1 (so TRACE values won't obscure the plot).
NOTE: These same WINDOW settings will work for a TI-89. Be sure to set "Use Freq and Categories" to Yes and specify the correct list of the frequencies.

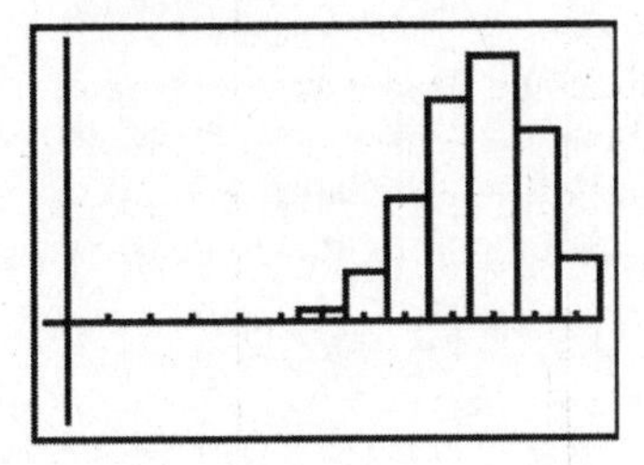

Press GRAPH to display the plot. The distribution is left-skewed with a visual center is around 9 Mexican-American jurors.

Mean, Variance and Standard Deviation

EXAMPLE Mexican-American Jurors: Use the probability distribution to find the mean number of Mexican-Americans in a jury of 12 members, the variance, and the standard deviation.

As before, I have the x values in L1 and the P(x) values in L2.

1-Var Stats L1,L2

Press STAT, arrow to CALC and press ENTER to select 1:1-Var Stats. We must tell the calculator first which list has the x values, and then which list has the probabilities. Press 2nd 1 , 2nd 2. Your command should look like the one at right. Press ENTER to see the results.

The value of the mean μ is given on this screen as $\bar{x}$ =9.598. The standard deviation σ is given by σx = 1.390106471, so the variance is $\sigma^2 = (1.3901)^2 = 1.932$. The calculator is smart enough to recognize that you are dealing with a probability distribution, so only gives σx and not a sample standard deviation; however, it has only one symbol for the mean, so you must know this is a μ (population) mean. Note that we also see n=1. Use this as a double-check that frequencies have been properly input. If n is *not* 1, there is either some rounding error, or an input error.

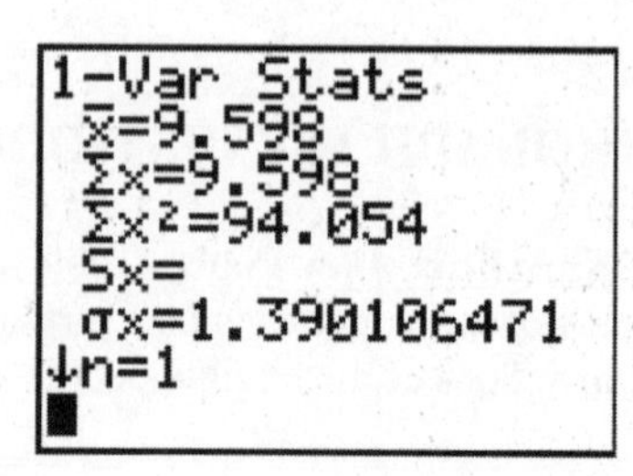

TI-89: The procedure is analogous, but set Use Freq and Categories to Yes, and name the list of probabilities as the Freq list.

Expected Value/Mean by the Formula $\sum[x * P(x)]$

EXAMPLE Mexican-American Jurors: Use the probability distribution to find the expected number or the mean number of Mexican-Americans on a jury of 12.

L1	L2	L3
0	0	------
1	0	
2	0	
3	0	
4	.001	
5	.003	
6	.016	

L3 =L1*L2

Again put the x values in L1 and the P(x) values in L2. This is partially shown at right.

With list name L3 highlighted, multiply L1 by L2 as shown in the bottom line. Press ENTER to perform the calculation.

sum(L3
9.598

Press 2nd MODE to Quit and return to the home screen. Then press 2nd STAT ▸ ▸ 5 to paste the sum function from the LIST, MATH submenu on the home screen. Then press 2nd 3 for L3. Press ENTER to find the sum of the entries in L3 which turns out to be 9.598 ≈ 9.6.

TI-89: Multiply the lists as above. To perform the sum operation, you can either place the cursor in a blank list within the statistics editor and use F3 to access the List menu, arrow to Math, then select menu option 5:sum(or on the home screen, press 2nd 5 (Math), select option 3:List, then select option 6:sum(.

BINOMIAL DISTRIBUTION

Binomial Probability Formula
$P(x) = nCx*p^x(1-p)^{n-x}$

EXAMPLE Jury Selection: Find the probability of getting exactly 7 Mexican-American jurors on a panel of 12 if the population is 80% Mexican-American. That is, find P(7) given that $n = 12$, $x = 7$, $p = 0.8$, and $q = 0.2$.

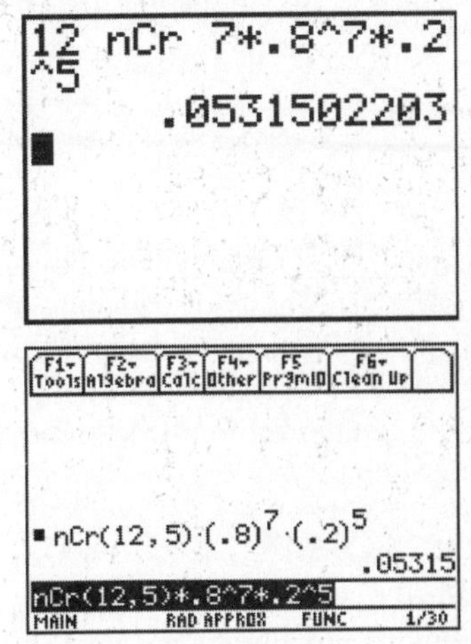

On a TI-83 or -84, type on the home screen to emulate what you see in the screen at right. You can find the `nCr` function in the `MATH, PRB` submenu (at [MATH] [◄] [3]). Press [ENTER] to calculate the probability as 0.0532.

On a TI-89, one must enter the `nCr` function first. Press [2nd][5] for `Math`, then [7] for the `Probability` menu, then [3] for `nCr`. Out of 12 jurors, we are interested in 7 being Mexican-American, so complete the combination by entering 12,7 and then closing the parentheses. Complete the calculation by multiplying by $*.8^7*.2^5$.

Built-in Binomial pdf and cdf Calculators

TI-83/84 Procedure

EXAMPLE Jury Selection: We will extend the above example and show how the TI calculators' built-in functions for the binomial distribution can be used for a variety of problem types.

(a) Find the complete probability distribution.
To get the complete table of values and their probabilities, we first recognize that the possible number of Mexican-American jurors on a panel of 12 is somewhere between 0 and 12. Fill list `L1` with the values 0 to 12. With the list name highlighted, press [2nd][STAT] (`LIST`), arrow to `OPS`, then select option `5:seq(`. Complete the command by typing X,X,0,12). Press [ENTER] to populate the list. To find the associated probabilities, move the cursor to highlight the name of list `L2`. Press [2nd] [VARS] (`DISTR`) and arrow to find binompdf. The actual location of this command will vary. It is menu option `Ø` on TI-83 calculators. If you have a TI-84 with operating system 2.30, it is menu option `A`. Then type 12 [,] 0.8 to specify that n = 12 and p = 0.8.

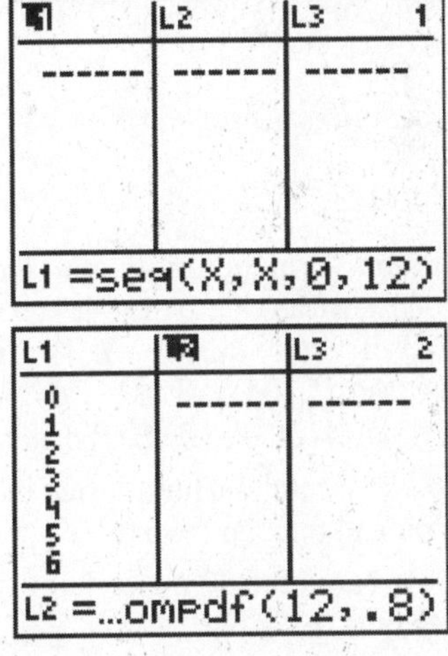

Press [ENTER] to calculate the probabilities. We can easily see the probabilities associated with different numbers of Mexican-Americans on a 12-member jury.

TI-89 Procedure:

In the Statistics Editor, press [F5] (`Distr`). Select option `B:Binomial Pdf`. In the dialog box, enter 12 for n, and 0.8 for p. Leave the box labeled `x` blank.

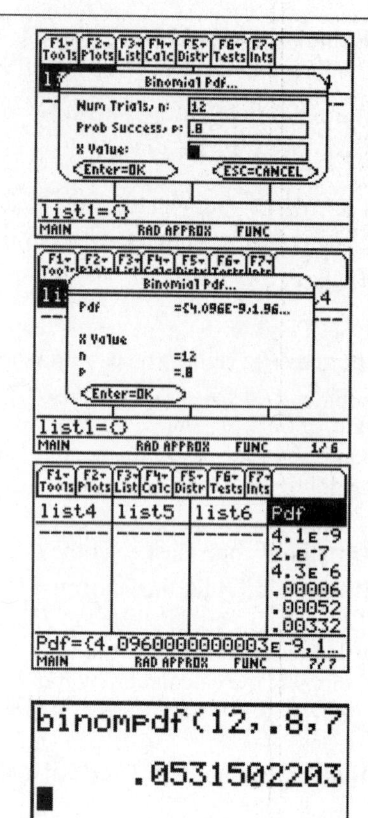

Pressing [ENTER] will display this results box.

Pressing [ENTER] again will display a new list labeled `Pdf` with the complete probability distribution. Please bear in mind that the first entry in the list is P(0), that is the probability of no Mexican-Americans on the jury. As you scroll down the list of probabilities, subtract one from the subscript in the list for the appropriate value of x.

(b) Find the probability of getting *exactly* 7 Mexican-Americans.

To find the probability at a single point, we use the `Binompdf` function. Press [2nd] [VARS], arrow to `Binompdf`, press [ENTER] to select it, then type 12 [,] .8 [,] 7 and then [ENTER] for the answer .0532 as seen in this screen. Of course, the answer agrees with what was seen on the previous page when we used the formula explicitly.

Note: pdf stands for probability density function

TI-89: Use `Binomial Pdf` as described above, except now specify the *x* value of 7. Pressing [ENTER] in the dialog box displays the results at right.

(c) Find the probability of *at most* 7 Mexican-Americans. (TI-83/84)

At most means "less than or equal to", so the probability of at most 7 is in fact P(0)+P(1)+P(2)+...+P(7). This can be found by using the built-in `binomcdf` (`Binomial Cdf` on an 89) function. This is right below `Binompdf` on the `DISTR` menu (option A, B, or C depending on calculator model. Hint: locate it by pressing [▲] to go to the bottom of the menu and then up from there.) With the command pasted on the home screen (or in the 89 dialog box), type 12,0.8,7 and press [ENTER]. The answer is 0.726.

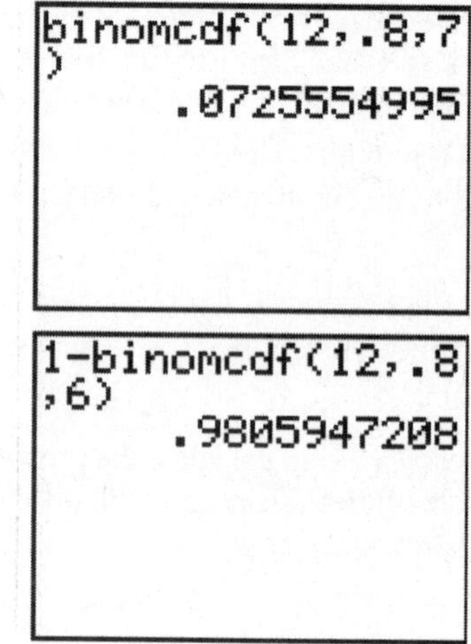

(d) Find the probability of *at least* 7 Mexican-Americans. (TI-83/84)

At least means "greater than or equal to". We want P(7)+P(8)+...+P(12). Since the Binomcdf function finds $P(X \leq k)$, we will find this as 1 - [P(0)+P(1)+P(2)+...+P(6)].

(e) *At most* and *at least* (TI-89)
The `Binomial Cdf` function of the TI-89 works differently from the other calculators. On its dialog box one specifies the `Lower Value` (the lowest x value to be included) and the `Upper Value` (the largest x value to be included). The calculation set up in this screen will find $P(X \leq 7)$. To find $P(X \geq 7)$, one specifies a lower value of 7 and an upper value of 12.

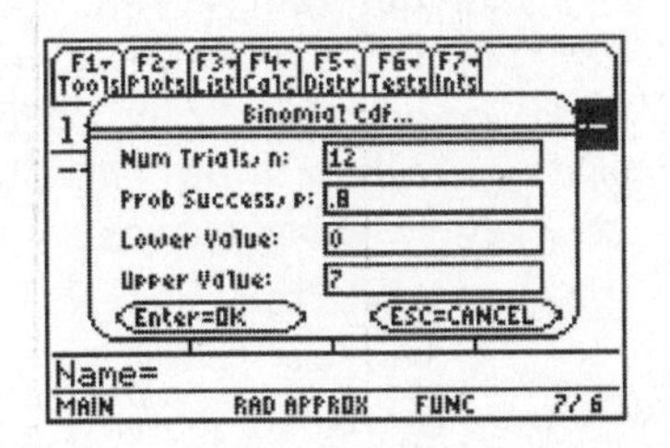

POISSON DISTRIBUTION

EXAMPLE World War II Bombs: (The problem below has been extended to answer more questions in order to better show off the capabilities of the TI calculators). In analyzing hits by V-1 buzz bombs in World War II, South London was subdivided into 576 regions, each with an area of 0.25 sq. km. A total of 535 bombs hit the combined area of 576 regions for an average of 535/576 = 0.929 hits per region. If a region is randomly selected, use the Poisson distribution to answer the following questions.

Poisson Probability Formula $P(x) = \mu^x e^{-\mu}/x!$

(a) Find the probability that the region was hit *exactly* twice.
Recalling that in this problem $\mu = 0.929$, and noting that for this question x = 2, we evaluate the above formula on the home screen as seen at right.
Note: to get e^(, one presses [2nd][LN] on an 83 or 84. On an 89, e^(is [◆][X]. Also remember that on an 83 or 84, factorials are menu choice 4 from [MATH], PRB; on an 89 they are choice 1 from the `Math`, `Probability` menu.

```
.929^2*e^(-.929)
/2!
          .1704283575
```

Built-in Poisson pdf and cdf Calculators

(a) Find the probability that the region was hit *exactly* twice.
We can use the built-in function `poissonpdf(` to find the probability of a Poisson variable being equal to some given value x. This function is located in the `DISTR` menu toward the bottom (actual location depends on calculator model). Paste on the function and fill in the rest. Press [ENTER] to see the same result as before. Note that the `poissonpdf(` function requires two inputs: μ and x.

```
.929^2*e^(-.929)
/2!
          .1704283575
poissonpdf(.929,
2)
          .1704283575
```

(b) Find the probability that the region was hit *at most* twice.
The built in function `poissoncdf(` can be used to find cumulative probabilities of a Poisson variable up to and including a given value x. This function is located immediately below poissonpdf on the `DIST` menu. Paste it on the main screen and fill in the rest. We see the cumulative probability is .9323. Note the inputs for the `poissoncdf` function were the same as for the pdf function, μ and x.

```
poissonpdf(.929,
2)
          .1704283575
poissoncdf(.929,
2)
          .9322839397
```

(c) Find the probability that the region was hit *at least* twice.
We need the probability of greater than or equal to 2 hits. This is $P(2)+P(3)+P(4)+\ldots = 1 - [P(0)+P(1)]$.
We see that we need 1 – poissoncdf(0.929,1). We type this on the home screen and [ENTER] for the answer .2381.

```
1-poissoncdf(.92
9,1)
          .2381444179
```

To find this probability on a TI-89, remember that on that model we specify the low end of interest and the high end of interest (strictly, infinity or ∞, but practically a large number will do). In this dialog box I have specified the `Upper Value` (high end of interest) as 1000., since practically speaking, there is no probability above here. After pressing [ENTER] we get the same answer.

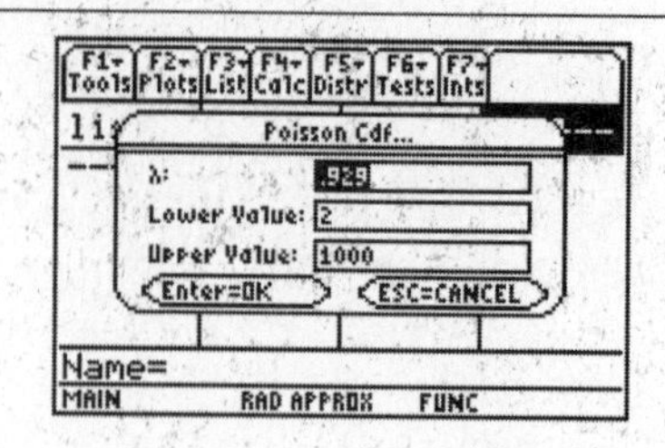

(d) Find the probability that the region was hit between 2 and 6 times, inclusive.

We see that we need P(2)+P(3)+P(4)+P(5)+P(6). We note that this will require us to find the value of `poissoncdf`(0.929,6) – `poissoncdf`(0.929,1). It is important to remember to close the parentheses after `poissoncdf(`0.929,6 on TI-83/84 calculators, otherwise it will not know that you are done with that `poissoncdf` and ready to move to another one.

We find this probability is 0.2381. On an 89, simply use poissoncdf and specify a `Lower Value` of 2 and an `Upper Value` of 6.

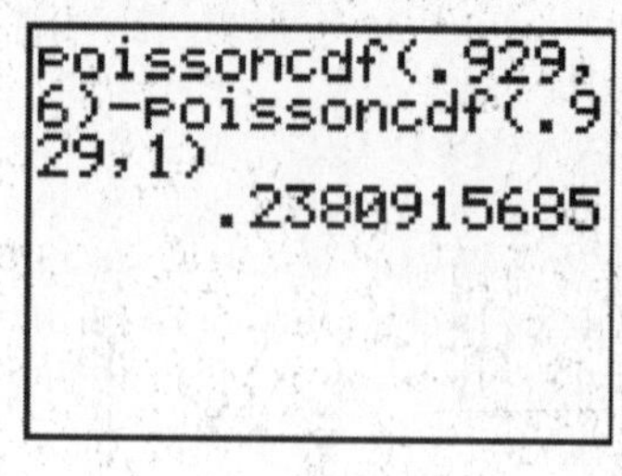

One important thing to note when dealing with discrete distributions that a question reading "between a and b" does NOT include the endpoints, since mathematically this would be $P(a < X < b)$. This is different from specifically including the endpoints, as above.

WHAT CAN GO WRONG?

Err: Domain?

This error is normally caused in these types of problems by specifying a probability as a number greater than 1 (in percent possibly instead of a decimal) or a value for n or x which is not an integer. Reenter the command giving p in decimal form. Pressing [ESC] will return you to the input screen to correct the error. This will also occur in older TI-83 calculators if n is too large in a binomial calculation; if that is the case, you need to use the normal approximation.

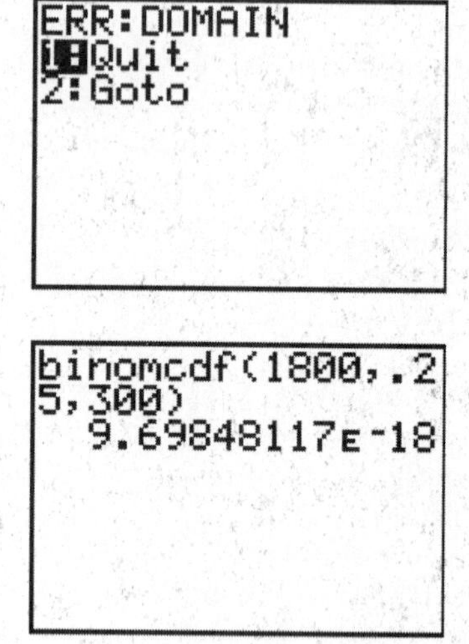

How can the probability be more than 1?

It can't. If a probability looks more than 1 on the first glance, check the right hand side. This value is 9.7×10^{-18} or seventeen zeros followed by the leading 9.

6 Normal Probability Distributions

In this chapter you will use your TI calculator to aid you in several types of problems dealing with the Normal Probability Distribution. First, you will learn to use the built-in function `normalcdf` to obtain probabilities associated with normal random variables. You will also learn to use the built-in function `invnorm` to obtain percentiles from a given normal distribution (that is, when given a probability value you will find the associated value of the normal variable). Both of these functions are on the DISTR menu at [2nd] [VARS] ([F5] `Distr` on an 89). Using them will alleviate the need to use the tables in the text. You will learn to do a simulation which illustrates the Central Limit Theorem and to approximate a binomial distribution with a normal distribution. You will learn how to obtain a *normal quantile plot*. This plot, described in the text, is used to determine if a given set of data might come from a population with a normal distribution. Finally, you will be introduced to a built-in feature which can generate random normal data for you to use in simulations of your own.

FINDING PROBABILITIES FOR A NORMAL RANDOM VARIABLE

We will use the built-in function `normalcdf(`. This function is in the DISTR menu as option 2. ([2nd] [VARS] [2] on TI-83 or 84 calculators). It requires 4 inputs. They are (a,b,μ,σ). Here the "a" and "b" denote two values between which you want the probability. As you might guess, μ and σ represent the mean and standard deviation of the normal variable. (If we do not put in values for μ and σ, the calculator assumes the distribution is *standard* normal and thus $\mu = 0$ and $\sigma = 1$). We will use the function on several examples to follow.

Standard Normal Probabilities

EXAMPLE Scientific Thermometers: The Precision Scientific Instrument company manufactures thermometers that are supposed to give readings of 0° C at the freezing point of water. Tests on a large sample of these instruments reveal that at the freezing point of water, some thermometers give readings above 0° C (denoted by positive numbers) and some give readings below 0° C (denoted by negative numbers. If we assume that the mean reading is 0° C, and the standard deviation of the readings is 1° C, and that the readings are normally distributed (have a bell-shaped distribution), then these thermometers have a *standard normal distribution,* because the mean is 1 and the standard deviation is 0.

(a) If a thermometer is randomly selected, find the probability that its reading at freezing will be *less than* 1.58°.

TI-83/84 Procedure:
Press [2nd] [VARS] [2] to paste `normalcdf(` onto the home screen. Then type –E99,1.58,0,1 and [ENTER]. To get the "E" press [2nd] [,].We find the answer is 94.3%; there is a 94.3% probability a random thermometer will read less than 1.58°.
Note: Normal Distributions are (theoretically) defined on the entire real line (-∞ , ∞). The value –E99 ($-1\text{x}10^{99}$) is a very large negative number (that the TI-83/84 calculators understand as -∞. We used it in the "a" position because we had no value at

which to start our probability area, and we wished to get all the area <u>below</u> 1.58. In practice, however, a very "large negative" number (say -999999 could be used for -∞. Note also as shown in the lower part of the screen that we did not have to specify the mean and standard deviation for a *standard* normal probability problem.

TI-89 Procedure:

In the Statistics/List Editor application, press F5 (`Dist`) and select menu option `4:Normal Cdf`. The dialog box asks for the `Lower Value` of interest (-∞ since we want all possible values less than 1.58), the `Upper Value` of interest (1.58), the mean (0), and the standard deviation (1) for the distribution of interest. To enter -∞, press (-) ◆ CATALOG.

Press ENTER for the results. We see there is a 94.3% chance a randomly selected thermometer of this type will read less than 1.58°C at the freezing point of water.

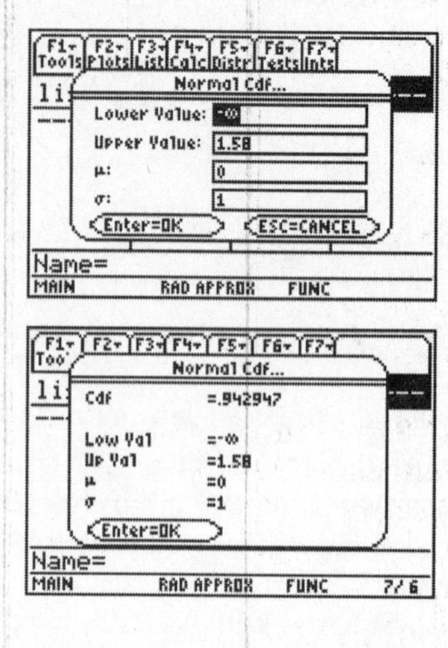

(b) Find the probability of a reading *above* –1.23°.

We proceed similarly to problem (a) but with different values for "a" and "b." We get 0.8907.

Note: This time we used E99 in the "b" position because it was a "greater than" problem and we had no value for the upper end of our probability area.

(c) Find the probability of a reading *between* –2° and 1.5°.

This is very straight-forward because in *between* problems, we are given values for inputs "a" and "b." We see the probability of a reading between -2° and 1.5° is about 91%.

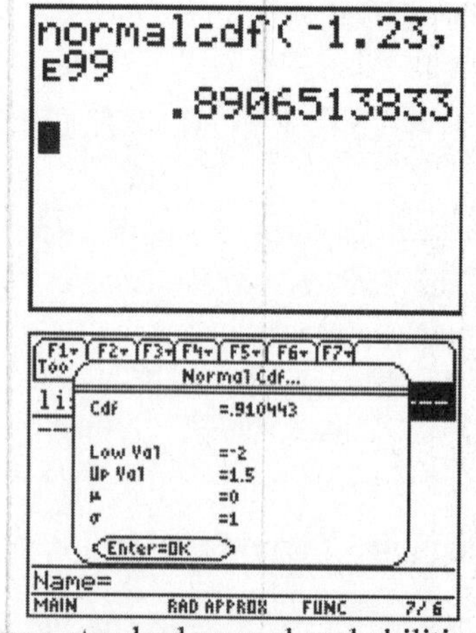

Non-Standard Normal Probabilities

There is really no difference in the TI-83 Plus' procedure for standard and non-standard normal probabilities using TI calculators except for the fact that in the non-standard normal problems the mean and standard deviation are required inputs for the `normalcdf` function.

EXAMPLE Weights of Water Taxi Passengers: In this chapter's problem, it was noted that the safe load for a water taxi was 3500 pounds. Further, this was based on an assumed average weight per passenger of 140 pounds. Data from the National Health and Nutrition Examination indicates that weights of men are normally distributed with mean 172 pounds and standard deviation 29 pounds. If one man is randomly selected, what is the chance he weighs less than 174 pounds (the value suggested by the National Transportation and Safety Board)?

We will first use `normalcdf` as before, but specifying the mean and standard deviation, since this distribution is not standard normal. We see that about 52.7% of adult men will weigh less than 174 pounds.

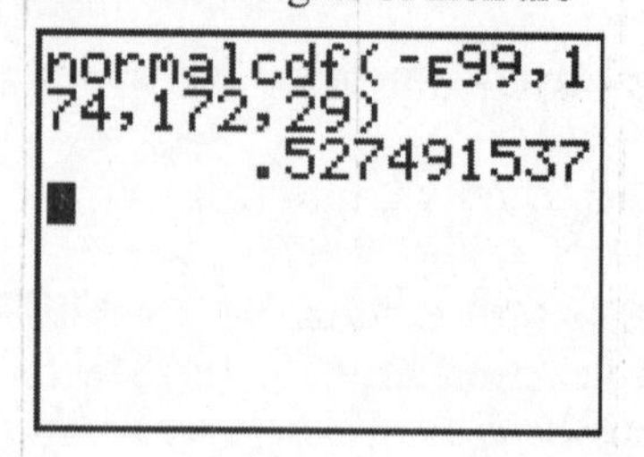

SKETCHING NORMAL AREAS

Your text suggests sketching the curve and shading the area corresponding to what is desired. This is very good advice, as it can help one decide if the values obtained from a calculation seem reasonable (Were all the parameters of the calculation input correctly?) TI calculators can also sketch normal curves and shade the desired area.

Caution: Before using this procedure, first turn off all `StatPlots`.

EXAMPLE: IQ Scores: A psychologist is designing an experiment to test the effectiveness of a new training program for airport security screeners. She wants to begin with a homogeneous group of subjects having IQ scores between 85 and 125. Given that IQ scores are normally distributed with a mean of 100 and a standard deviation of 15, what percentage of people have IQ scores between 85 and 125?

TI-83/84 Procedure:
Setting the WINDOW.
We have used the following criteria to set our window values:
`Xmin` = $\mu - 3\sigma = 100 - 3(15) = 55$
`Xmax` = $\mu + 3\sigma = 100 + 3(15) = 145$
`Xscl` and `Yscl` = 0
`Ymin` = $-.1/\sigma = -.1/15 = -0.01$
`Ymax` = $0.4/\sigma = 0.4/15 = 0.03$

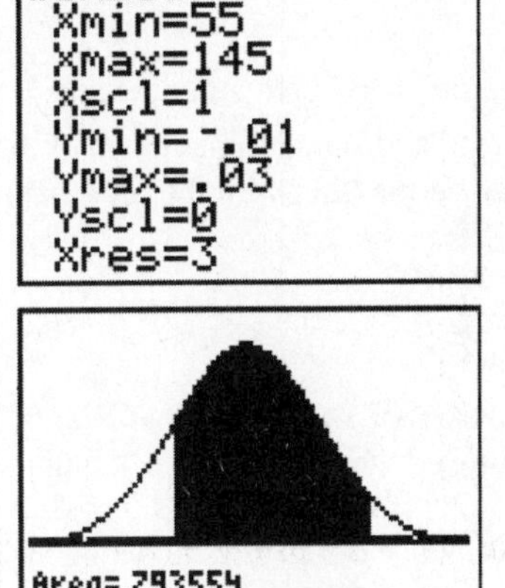

Getting the Graph: Press [2nd] [VARS] [▶] [1] to get `ShadeNorm(` pasted on your home screen. Fill in the values of a, b, μ and σ as usual. Press [ENTER] for the shaded. We can see the Area =.97725. This area is equivalent to the probability we were seeking.
Caution: If repeatedly using this procedure, the normal curve may become totally shaded! Between uses, press [2nd][PRGM] (Draw) and press [ENTER] to select `1:ClrDraw`. About 79.4% of all people should have IQ scores between 85 and 125.

TI-89 Procedure:
There is an option in the TI-89 dialog box which makes setting the window automatic. Press [F5] (`Distr`) and press ⊙ to expand the `1:Shade` submenu. Press [ENTER] again to select `1:Shade Normal`. The input dialog box looks just like that for `Normal Cdf`, with the exception of one option at the bottom: `Auto-Scale`. Press ⊙ to expand the choices, and set this option to Yes. Press [ENTER] to execute the command.

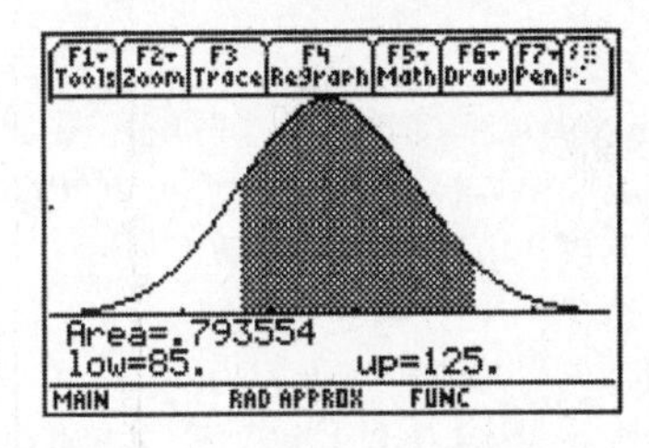

FINDING VALUES OF A NORMAL R. V. (INVERSE NORMAL PROBLEMS)

The InvNorm Function

EXAMPLE Scientific Thermometers: Using the same thermometers as earlier, find the temperature corresponding to P_{95}, the 95th percentile. That is, find the temperature separating the bottom 95% from the top 5%.

TI-83/84 Procedure:

Press [2nd] [VARS] [3] to paste `invnorm(` onto your home screen. This function needs three inputs: p, μ, and σ. Here p represents the percentile we desire to know (the area *to the left* of the desired point on the distribution). So for this problem we input 0.95. (We need not input μ, and σ since the thermometers are standard normal) Press [ENTER]. We see that 95% of these thermometers will read 1.64° or less at the freezing point of water.

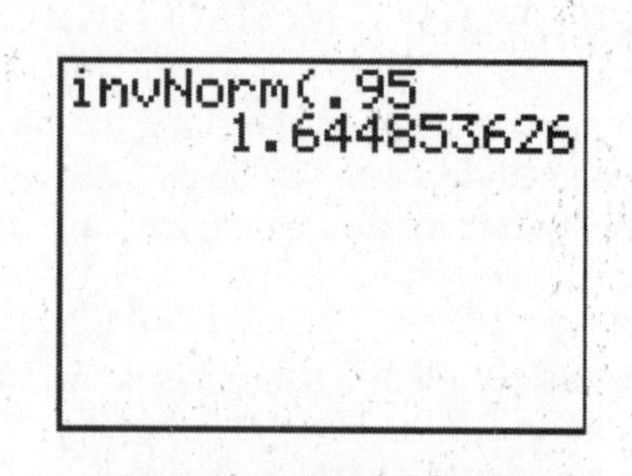

TI-89 Procedure:

The TI-89 has the capability of computing inverses for several distributions. In the Statistics List Editor, press [F5] (`Distr`) then arrow to `2:Inverse` and press (▶) to expand the submenu. Press [ENTER] to select `1:Inverse Normal`.

In the dialog box, enter the area to the left of the point desired, then the mean and standard deviation of the distribution. Pressing [ENTER] displays the output box, with the result value of 1.64°.

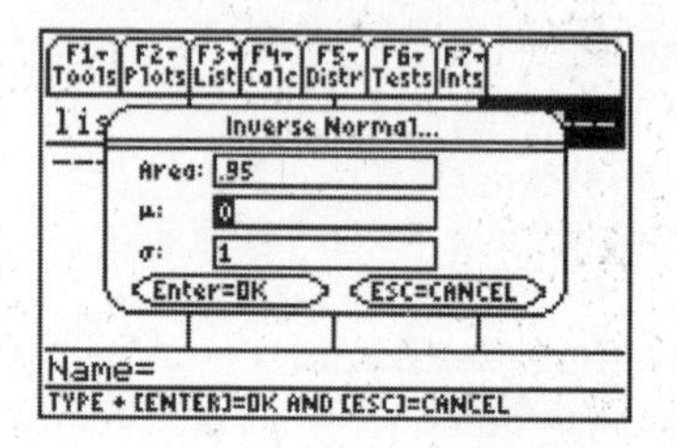

EXAMPLE: Weights of Water Taxi Passengers: A previous example showed that about 52.7% of all men have weights less than 174 pounds. What weight separates the lightest 99.5% from the heaviest 0.5%? Again, assume that weights of men are normally distributed with mean 172 pounds and standard deviation 29 pounds.

From the `DISTR` menu ([2nd][VARS]) paste `invnorm(` onto your home screen. This function needs three inputs in this case: p, μ, and σ. Here p represents the percentile we desire to know. So for this problem we input 0.995,172,29 and [ENTER]. We see that a weight of 246.7 pounds separates the 99-1/2% from the top half-percent.

Note: Make sure to use the decimal value of the desired percentile (for example 0.98 and not 98).

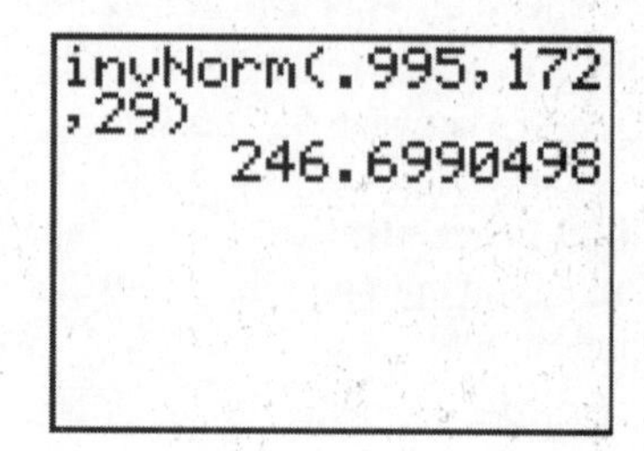

EXAMPLE CD Players: When deciding placement of a CD player in a car, designers need to consider the forward grip reach of the driver (otherwise, there might be an accident as the driver is changing U2 for Barry Manilow). Designers have decided to place the CD so that 95% of all women can reach it. Women (based on anthropomorphic survey data from Gordon, Churchill, et al) have forward grip reaches that are normally distributed with mean 27.0 in and standard deviation of 1.3 in. Find the forward grip length that separates the longest 95% from the others.

In the screen at right, we see the answer is about 24.9 inches. The point to be careful on with this problem is that we want to separate the *longest* 95% from the *shortest* 5%. Keeping in mind that `invNorm` works with area to the left of the desired point, we enter .05 (5%) as the area, not .95 (95%.)

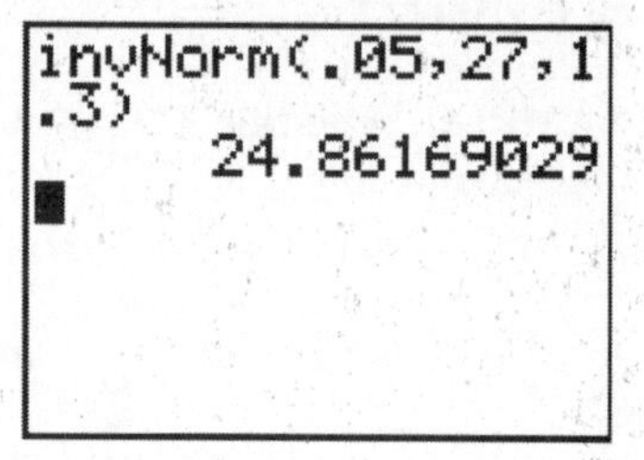

CENTRAL LIMIT THEOREM

As the sample size increases the sampling distribution of the sample means approaches a normal distribution.

We perform a simulation of the example in the text which looked at the last 4 digits of 50 randomly selected social security numbers. In this simulation, you will generate 50 sets of 4 digit numbers on the TI-83 Plus and look at the distribution of the numbers selected and the distribution of their means.

Set the seed, so your random number generator will duplicate the results shown here. Press 4321 [STO▸] [MATH] [◄] [1] [ENTER].

Set up 4 lists of 50 random integers between 0 and 9.
This is partially shown. The keystrokes are as follows:

[MATH] [◄] [5] 0 [,] 9 [,] 50 [STO▸] [2nd] [1] [ENTER]
[MATH] [◄] [5] 0 [,] 9 [,] 50 [STO▸] [2nd] [2] [ENTER]
[MATH] [◄] [5] 0 [,] 9 [,] 50 [STO▸] [2nd] [3] [ENTER]
[MATH] [◄] [5] 0 [,] 9 [,] 50 [STO▸] [2nd] [4] [ENTER]

```
4321→rand
                4321
randInt(0,9,50)→
L1
{5 3 7 5 1 9 9 …
randInt(0,9,50)→
L2
```

Note: Remember you can use [2nd] [ENTER] to recall the last entry and make this sequence of steps easier. See page 7.

TI-89 Note: Do this in the Statistics/List Editor. Store the seed using option `A:RandSeed` from the [F4] (`Calc`), `4:Probability` menu. To obtain the lists, highlight a list name, then use menu option `5:RandInt` from the Probability menu. The parameters of the command are just as above.

Find the mean of each of the 50 rows using the following: `(L1+L2+L3+L4)/4` [STO▸] [2nd][5] [ENTER]. On an 89, highlight the name `list5` then enter the formula.

Use the Statistics Editor to view your lists. Note the means in L5 are tending toward the middle of the distribution.

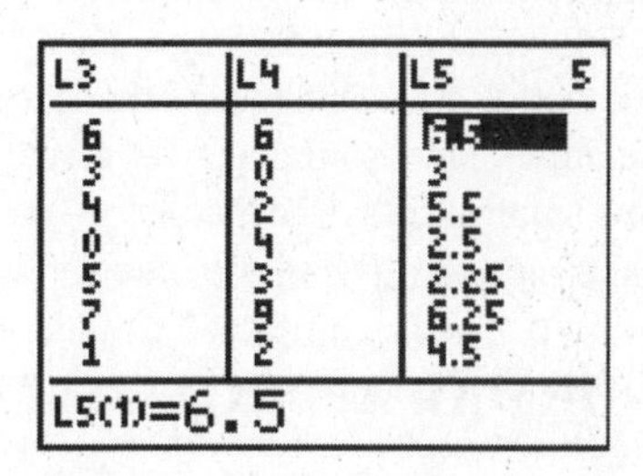

Set `Plot1` to get a histogram of the data in `L1`. For this graph, I have set `Xmin` to -.5, `Xmax` to 9.5, and `Xscl` to 1 since we want to see a bar for each digit. Note that the data in `L1` are fairly uniformly distributed between 0 and 9. The other lists would behave similarly.

```
P1:L1
min=2.5
max<3.5        n=8
```

Now change the setup of `Plot1` to graph the means stored in `L5`. Press [GRAPH] to display the histogram using the same settings as the individual lists. Note that this histogram is much more *normally* distributed than that in the previous screen. This is as predicted by the Central Limit Theorem.

P1:L5
min=4.5
max<5.5
n=14

EXAMPLE Water Taxi Safety: Assume (based on the National Health and Nutrition Examination) that the population of men have weights that are normally distributed with a mean of 172 lb and a standard deviation of 29 lb.

(a) Find the probability that if an individual man is randomly selected, his weight will be greater than 174 pounds. The probability is 0.4725.

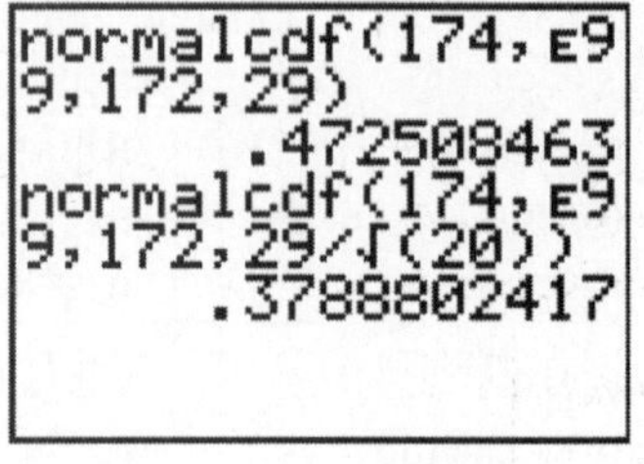

(b) Find the probability that 20 randomly selected men will have a mean weight that is greater than 174 pounds (so the total weight exceeds the safe limit of 3500 pounds for the water taxi.) The probability is 0.3789. Note that the only thing that has changed is the standard error which is now $29/\sqrt{20}$. You can enter it explicitly as I have in this example, or perform the calculation first ($29/\sqrt{20} = 6.4846$) but the temptation to round too much may be tempting! In any case, we can see that the chance of overloading the water taxi with 20 randomly selected men is too high. There should be a lower person limit for safety's sake.

Note: In the above, our answers differ slightly from those in the text because we did not use the tables. The calculator solutions are actually more precise because they do not suffer from round-off error.

EXAMPLE: Body Temperatures: If we assume that humans have a mean body temperature of 98.6°F with standard deviation 0.62 °F, what is the chance of getting a mean reading of 98.2 °F or lower from a random sample of 106 people?

We don't know the shape of the original distribution, but with the large sample size, the sample mean should have a normal distribution, due to the Central Limit Theorem. Therefore, we can answer the question using normalcdf as at right. We need to be careful reading the answer. At first glance, the probability might look larger than 1 (1.552). *But, probabilities can't be larger than 1!* Look at the right-hand portion of the answer: `E-11`. This means the leading digit (1) occurs in the eleventh decimal place, so the answer is really 0.000000000016. A probability this small usually means that given an observation (the sample mean actually seen) something is wrong with an assumption (here, the mean body temperature is probably less than the 98.6 °F usually assumed.)

NORMAL APPROXIMATION OF A BINOMIAL DISTRIBUTION

If np $\geq$ 5 and n(1-p) $\geq$ 5, then the binomial random variable is approximately normally distributed with the mean and standard deviation given by $\mu = np$ and $\sigma = \sqrt{np(1-p)}$. **(Actually, the criterion varies from textbook to textbook, but this is what Triola uses.)**

EXAMPLE Passenger Load on a Boeing 767-300: Find the probability that among 213 randomly selected passengers there are at least 122 men (too many men might result in an unsafe situation and the load would have to be adjusted). Assume that the population of passengers consists of an equal number of men and women.

First, we check to see if we can use the normal approximation. We see that $np = 213(0.50) = 106.5 > 5$ and that $n(1-p) = 213(0.50) = 106.5 > 5$. So we can use the normal approximation to the binomial. We can approximate $\mu = np$ and $\sigma = \sqrt{np(1-p)}$ by $\mu = 106.5$ and

$\sigma = \sqrt{213(0.5)(1-0.5)} = 7.29726$

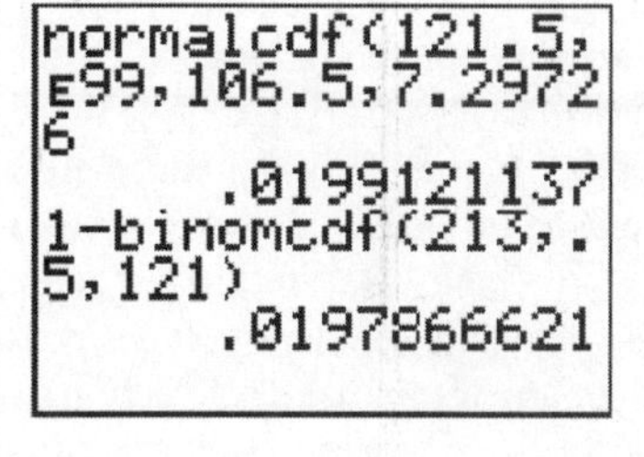

Using the continuity correction to find P(X > 122), we need to start at 121.5 and continue on to the right. We obtain a probability of 0.0199. (Your answer may vary slightly depending on the number of decimal places used for the standard deviation). At the bottom of the screen shown, we check the answer by using the `binomcdf` function discussed in Ch. 5. We find the actual probability is 0.0198. We see that the approximation of .0199 was quite accurate. It should be a rare case that the plane's load would have to be adjusted due to too many men.

EXAMPLE Internet Use: A recent survey showed that among 2013 randomly selected adults, 1358 (or 67.5%) stated that they are Internet users (data from Pew Research Center). If the proportion of all adults using the Internet is actually 2/3, find the probability that a random ample of 2013 adults will result in exactly 1358 users.

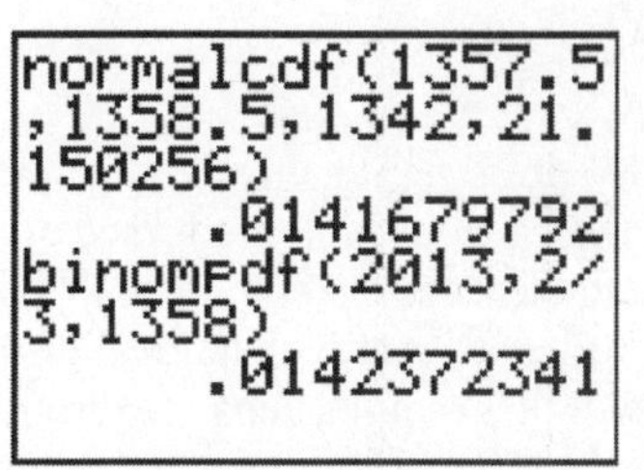

Again, we check to see if we can use the normal approximation. The mean is $np = 2013*2/3 = 1342 > 5$ and we also have $n(1-p) = 2013*1/3 = 671 > 5$. We need to find the standard deviation $\sigma = \sqrt{2013(2/3)(1-2/3)} = 21.150256$. According to the continuity correction, we will consider "exactly 1358 Internet users" to be anywhere between 1357.5 and 1358.5 Internet users. Notice we get results using the approximation and the binomial (exact) model that are extremely close. There is about a 1.4% chance to find exactly 1358 Internet users among 2013 randomly selected adults (actually, finding exactly *any* single number of Internet users in a sample this size would be small).

NORMAL QUANTILE PLOTS

EXAMPLE Heights of Men: Data Set 1 of Appendix B has information on health exam results for 40 men, including their heights (measured to the nearest one-tenth inch). We want to determine whether or not these data indicate they are from a normal population.

TI-83/84 Procedure:
Get the data set into `L1`. You can type it, transfer it from another calculator, or transfer it from Data Apps (see Appendix, page 107).

Set up `Plot1` as at right. Note that the normal quantile plot is the <u>last</u> plot type. Make sure you choose `X` as the data axis.

Press [ZOOM] [9] for the plot. Note the normal quantile plot is fairly linear and thus these men's heights seem normally distributed. There is an indication on the right edge of the graph that there might be an outlier (the last point is separated from the rest). In fact, a modified box plot does indicate this individual is unusually tall (76.2" or 6'4.2").

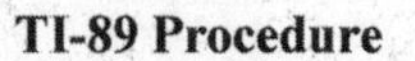

TI-89 Procedure

Producing normal quantile plots on a TI-89 calculator is somewhat more involved than on TI-83's and -84's.

With the data entered in `list1`, press [F2] (`Plots`) and select choice `2:Norm Prob Plot`. Use the right and down arrows to select a plot number to use (the calculator will default to a minimum one more than the currently defined plots). Enter the list name containing the data (use [2nd][–] = [VAR-LINK] to find the list name), select X as the data axis, and choose whatever mark you desire for each data point (use the right and down arrows to make a selection). Notice the calculator will create a new list for the z-scores of the data values.

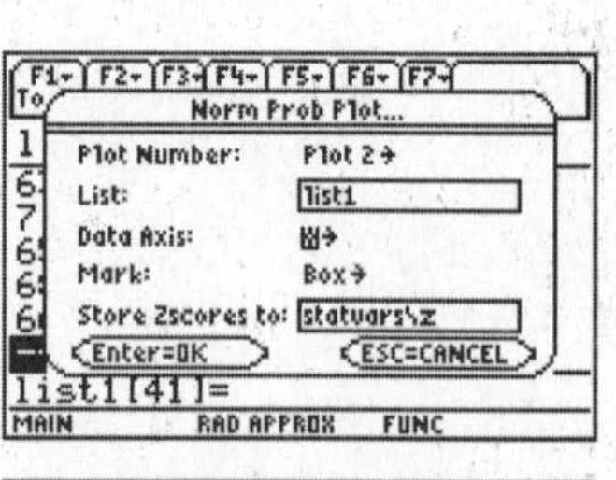

After pressing [ENTER], notice the new list created in the list editor.

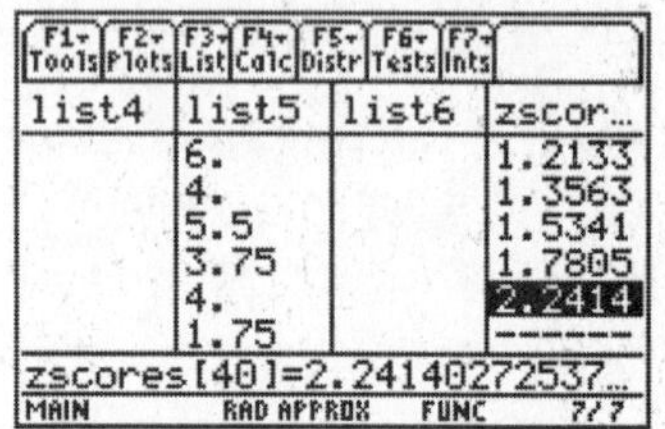

Now, press [F2] again and select option `1:Plot Setup`. Make sure any extraneous plots are turned off (Move the highlight to them and press [F4] to clear the check). Press [F5] (`Zoom Data`) to display the plot. This plot shows a relatively normal distribution with the possibility of a high end outlier (a point separated from the rest).

To clear the z-scores list, return to the Statistics Editor, and press [F1] (`Tools`) and select option `3:Setup Editor`. Leave the box of "`Lists to View`" blank.

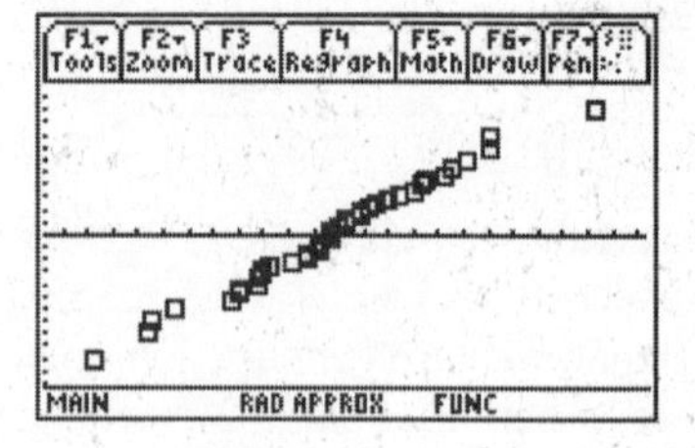

GENERATING RANDOM VALUES FROM A NORMAL DISTRIBUTION

If you ever wish to do simulations which require values from a normal distribution, TI calculators have a built-in function which will allow you to generate the random values you will need. For example, suppose you need 25 random values from a *standard* normal distribution (mean = 0, standard deviation =1).

Set your seed to 543, so your answers will duplicate those seen here. 543 [STO▸] [MATH] [◂] [1] [ENTER] (On an 89, be sure to use option `A:Rand Seed` from the [F4] (`Calc`, `4:Probability` menu.)

Next, select option `6:randNorm` from the `MATH`, `Prb` submenu by pressing [MATH] [◂] [6]. You next need to specify the mean, the standard deviation and how many values you wish to generate. Fill in these as at right to generate 25 values from the standard normal and store them in `L1`. Press [ENTER]. (On an 89, this is option 6 from the [F4] `Probability` menu. Do this in the Statistics List Editor with the name of the destination list highlighted.)

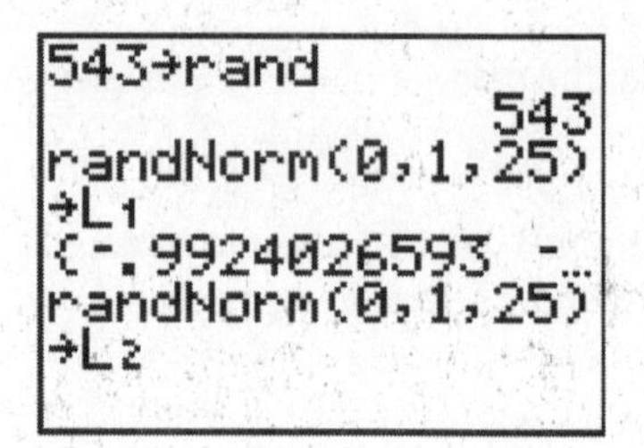

Repeat the `randNorm` command, this time storing in `L2`. Press [ENTER].

Sort the data sets in L1 and L2, so you can look at the data as at right. Note that the sets are different yet similar in distribution.

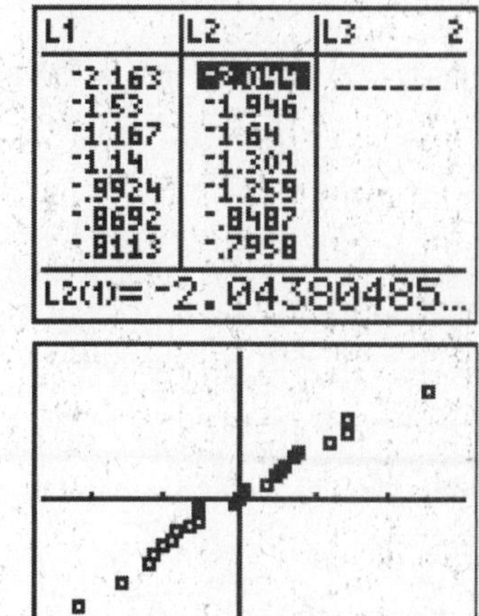

Now setup Plot1 for a normal quantile plot using the data stored in L1 as described above. Press [ZOOM] [9] to display the plot.

You can also change the StatPlot setup to plot the values from L2. You can see that the data generated in both runs is fairly normally distributed. You can try different sizes of data sets and different parameters (μ and σ) to get a feeling for normal distributions.

WHAT CAN GO WRONG?

Why is my curve all black?

In this curve, the graph indicates more than half of the area is of interest between z-scores of –3 and -.25; the message at the bottom says the area is 40%. This is a result of having failed to clear the drawing between commands. Press [2nd][PRGM] then [ENTER] to clear the drawing, then reexecute the command.

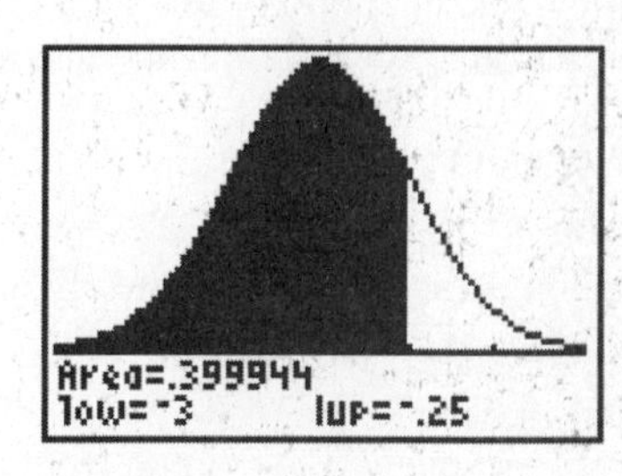

How can the probability be more than 1?

It can't. If the results look like the probability is more than one, check the right side of the result for an exponent. Here it is –4. That means the leading 2 is really in the fourth decimal place, so the probability is 0.0002. The chance a variable is more than 3.5 standard deviations above the mean (this would be a thermometer reading 3.5°C or more at freezing) is about 0.02%.

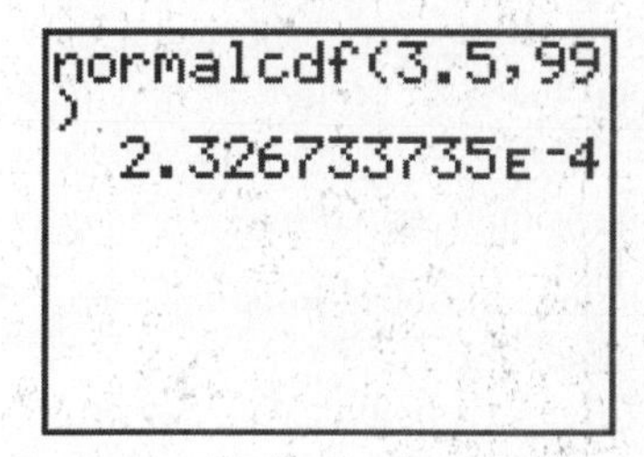

How can the probability be negative?

It can't. The low and high ends of the area of interest have been entered in the wrong order. As the calculator does a numerical integration to find the answer, it doesn't care. You should.

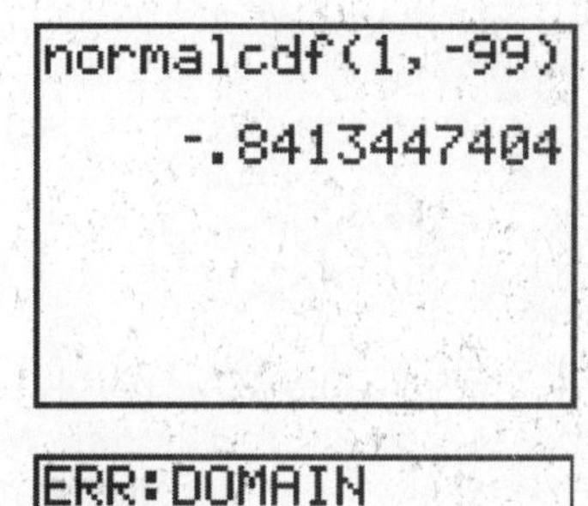

What's Err: Domain?

This message comes as a result of having entered the invNorm command with parameter 90. (You wanted to find the value that separates the top 10% of IQ scores, so 90% of the area is to the left of the desired value.) The percentage must be entered as a decimal number. Reenter the command with parameter .90.

```
ERR:DOMAIN
1:Quit
2:Goto
```

7 Estimates and Sample Size

In this chapter you will learn how your TI calculator can aid you in estimating population parameters using the results of a single random sample. In this work, you will primarily use some of the options on the STAT TESTS menu (Ints on an 89). You will learn how to estimate population proportions, means and variances as well as how to estimate the sizes of the samples you will need in each setting.

As usual, the first time a significant difference between the TI-83/84 and TI-89 calculators is encountered, we present both procedures. After an initial run-through with a confidence interval, the differences are clearly apparent.

ESTIMATING A POPULATION PROPORTION

EXAMPLE Touch Therapy Success Rate: In the Chapter Problem, touch therapists participated in 280 trials of their ability to sense a human energy field. They were correct 123 times. We wish to estimate the true population proportion. Find the margin of error E that corresponds to a 95% confidence level for the proportion. Then find the 95% confidence interval.

TI-83/84 Procedure

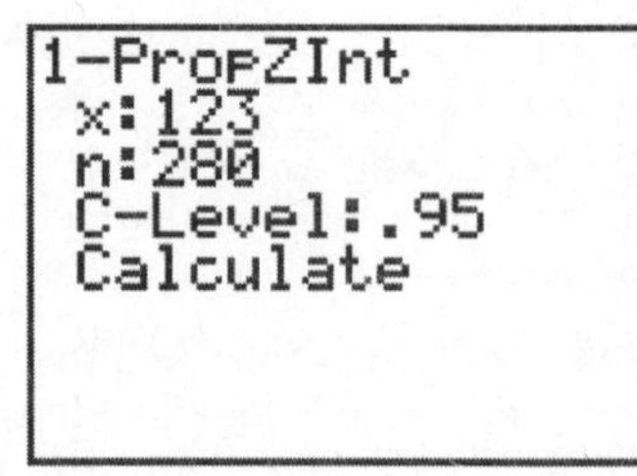

Press [STAT], arrow to TESTS, and arrow down to select menu and choose option A (actually, it is more efficient to use the *up* arrow, or press [ALPHA][MATH] = A). You will see a screen like that at right. Note that you need 3 inputs: x, n and C-level. In this problem it is obvious that n = 280 and C-level = .95. It is important to understand that the x which is needed is the number of responses with the characteristic of interest (not the percentage or proportion). Sometimes you are given x, but sometimes you are given the sample proportion $\hat{p}$ .To find x when you have been given the sample proportion $\hat{p}$ you must use the formula $x = n\hat{p}$ and round to the nearest integer. You can actually do the multiplication on the input line for x – press [ENTER] to do it, then arrow back up and do any needed rounding.

Note: The value of x must be an integer or else your TI calculator will have a domain error.

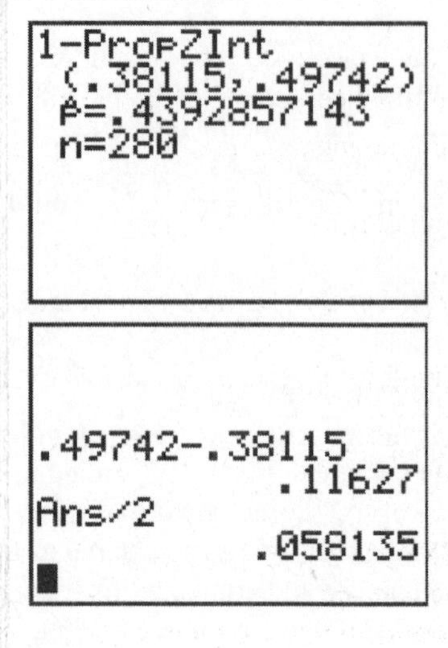

Highlight the word Calculate and press [ENTER] to see the results. The interval is given in parentheses. It is (0.38115, 0.49742). We normally report the interval only to the nearest 1/10th of a percent; here we would report our 95% confidence interval as 0.381 to 0.497, or 38.1% to 49.7%. Based on this study, tough therapists can correctly identify the electrical field between 38.1% of 49.7% of the time, with 95% confidence.

If you wish to find the margin of error E which was used to calculate this interval, you must find the difference of the two endpoints of the interval and divide by 2. We find that the margin of error was E = .058.

TI-89 Procedure

From the Statistics List Editor, press [2nd][F2] = [F7] (`Ints`). `1-PropZInt` is option 5 on this menu. After selecting the menu option (either press [5] or scroll down to highlight the option and press [ENTER]) you will seen an input dialog box like the one at right.

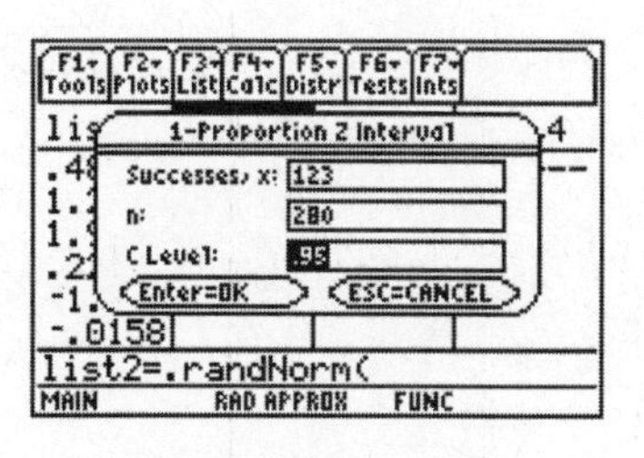

Input the number of successes, `x`, (this must be an integer), the number of trials, `n`, and the desired Confidence Level. If you are given the sample results as a proportion (percentage) of successes, multiply by `n` and round to find `x`.

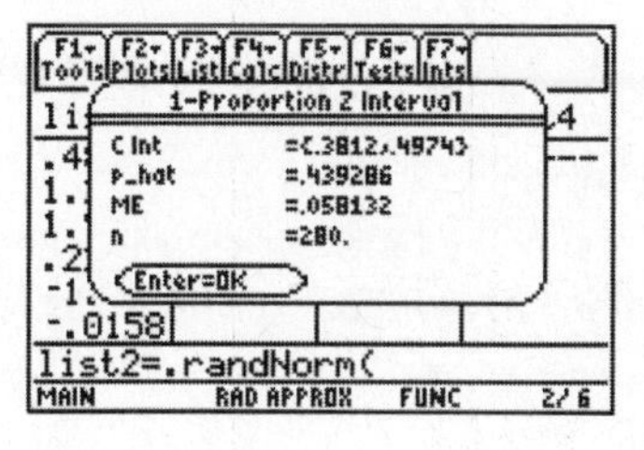

Press [ENTER] to display the results. Notice that the TI-89 gives the margin of error (`ME`) as well as the ends of the interval.

Home Screen Calculation for Proportion Estimates

Using the same example, you could calculate the margin of error and the confidence interval for the population proportion on your home screen.

You must find the value of $Z_{\alpha/2}$ for a 95% confidence level. To find this value, calculate as follows: $\alpha = 1 - 0.95 = 0.05$, $\alpha/2 = 0.05/2 = 0.025$. This means we need the value which separates the top 2.5% from the bottom 97.5% of the standard normal distribution (or the bottom 2.5% from the top 97.5% because of symmetry). We use the `invNorm` function as described in Chapter 6. We see the value needed is 1.96.

Now to calculate the value of the margin of error E use the formula $E = z_{\alpha/2}\sqrt{\frac{\hat{p}\hat{q}}{n}}$. We know from the problem that $\hat{p} = 0.44$, $\hat{q} = 0.56$ and $n = 280$. This is shown at the top of the screen at right. We have stored the margin of error as `E`.

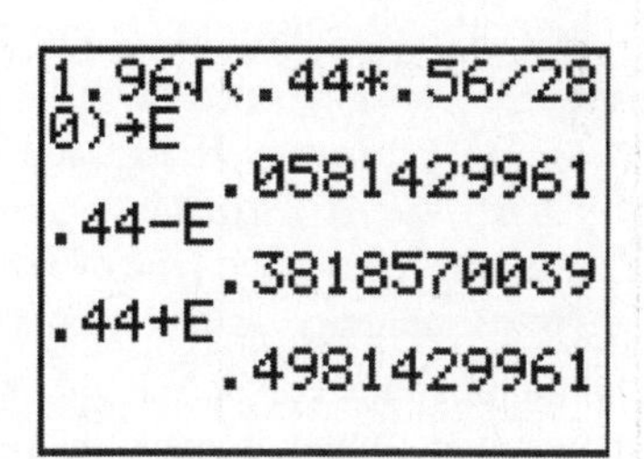

Now we continue to calculate the lower and upper endpoints of the confidence interval by taking our sample proportion $\hat{p}$=0.44 and subtracting and adding `E`. We obtain nearly the same two interval endpoint values found when we used `1PropZInt` (the differences are due to rounding).

Determining Sample Size Required to Estimate the Population Proportion

The formula for sample size is not built into TI calculators. We calculate it on the home screen. It is $n = \frac{[z_{\alpha/2}]^2\,\hat{p}\hat{q}}{E^2}$, where $\hat{p} = 0.5$ if no other estimate is available.

EXAMPLE Sample Size for E-Mail Survey: Suppose a sociologist wants to determine the current percentage of U.S. households using e-mail. How many households must be surveyed in order to be 90% confident that the sample percentage is within four percentage points of the true percentage for the nation?

a) Use the result from an earlier study which said 16.9% of U.S. households used e-mail.
b) Assume that we have no prior information suggesting a possible value of $\hat{p}$.

```
invNorm(.05
    -1.644853626
```

First find the critical z-value for 90% confidence. As explained on the previous page we use `InvNorm` with $\alpha/2 = 0.05$. We find the value is 1.645. Ignore the negative – the normal distribution is symmetric, and z will be squared.

```
1.645²*.169*.831
/.04²
        237.5196531
1.645²*.5*.5/.04
²
        422.8164063
```

We show the calculations using both values for $\hat{p}$. We find that the sample size needs to be 238 if we use the prior estimate for $\hat{p}$ and 423 if we do not have a prior estimate and must use $\hat{p} = 0.5$. Notice that having a "guessed" starting value can save lots of work!

ESTIMATING A POPULATION MEAN: σ KNOWN

The formula for the confidence interval used to estimate the mean of a population when the standard deviation is known is built into the calculator. It is option 7 on the `STAT TESTS` menu on the TI-83/84 (option 1 on the `Ints` menu on a TI-89) and is called `ZInterval`. This built-in function can be used with either raw data or summary statistics. The interval could also be calculated on the home screen in a similar manner as was shown for the confidence interval for a population proportion.

EXAMPLE Pulse Rates of Females: For the sample of pulse rates of women in Data Set 1 of Appendix B, we have $n = 40$ and $\bar{x} = 76.3$, and the sample is a simple random sample. We will assume that σ is known to be 12.5. Using a 0.95 confidence level, find the margin of error and the confidence interval.

ZInterval with Summary Statistics

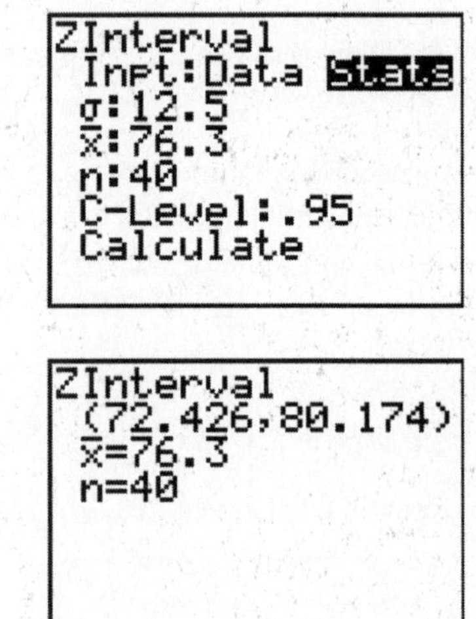

Press STAT, arrow to `TESTS` and select option `7:ZInterval` (option 1 on the TI-89 `Ints` menu) to get a screen similar to the one at right. Your first choice is the type of `Input` you will use. For this problem you have summary statistics (not raw data entered in a list), so you should move the cursor to highlight `Stats` and press ENTER. This choice is presented on the TI-89 before the input dialog box is shown. Now fill in all the other information which was given in your problem.

Highlight the word "Calculate" and press ENTER to display the results. Based on this sample, we are 95% confident the average pulse of a woman is between 72.4 and 80.2 beats per minute.

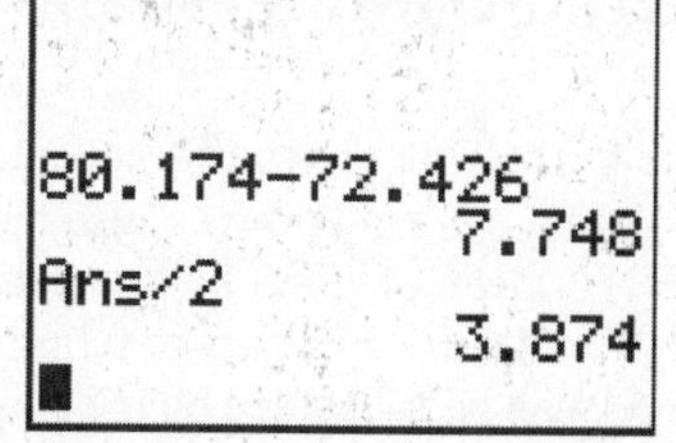

To retrieve the margin of error that was used to build the interval, we again must find the difference between the upper and lower endpoints of the interval and divide this difference by 2. The margin of error was 3.9 beats per minute. (The TI-89 again gives the margin of error (ME) explicitly in its results).

ZInterval with Raw Data

Using the Pulse Rate example above, assume you do not yet have the sample statistics, but you have the 40 body temperature values in Data Set 1 of Appendix B.

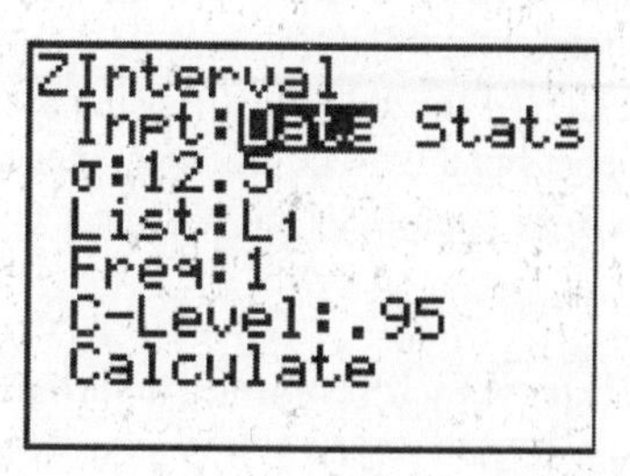

Put the data into list L1. (Do this by typing it in or transferring from another source.)

Press STAT, arrow to TESTS and select ZInterval. Your first choice is the type of Input you will use. For this problem you have raw data, so you should move the cursor to highlight Data and press ENTER. Now fill in all the other information. Note you must specify the standard deviation because it is supposed to be known and also the list where you have stored your data.

Highlight the word Calculate and press ENTER. You should see the same results as above since the outcome of the confidence interval is the same as when it was done with summary statistics.

Home Screen Calculations for Mean Estimates

To calculate the margin of error and confidence interval for the Pulse Rates example on the home screen, we would first need to know the value of $Z_{\alpha/2}$ for a 95% confidence level. We already found this in to be 1.96.

```
1.96*12.5/√(40)→E
                3.873790134
76.3-E
                72.42620987
76.3+E
                80.17379013
```

Here, we used the formula $\bar{x} \pm E$ where $E = z_{\alpha/2} \dfrac{\sigma}{\sqrt{n}}$. We first calculated E, stored it as E and then used it to find the endpoints of the confidence interval for the mean pulse rate. Again, the interval is (72.4, 80.2).

Determining Sample Size Required to Estimate μ

TI calculators do not have a built-in function to calculate sample size. The formula can be calculated on the home screen. The procedure is similar to the previous example of a sample size calculation for a proportion problem, but the formula is different: $n = \left[\dfrac{z_{\alpha/2}\sigma}{E}\right]^2$.

EXAMPLE: IQ Scores of Statistics Professors: Assume that we want to estimate the mean IQ score for the population of statistics professors. How many statistics professors must be randomly selected for IQ tests if we want 95% confidence that the sample mean is within 2 IQ points of the population mean (assuming $\sigma = 15$)?

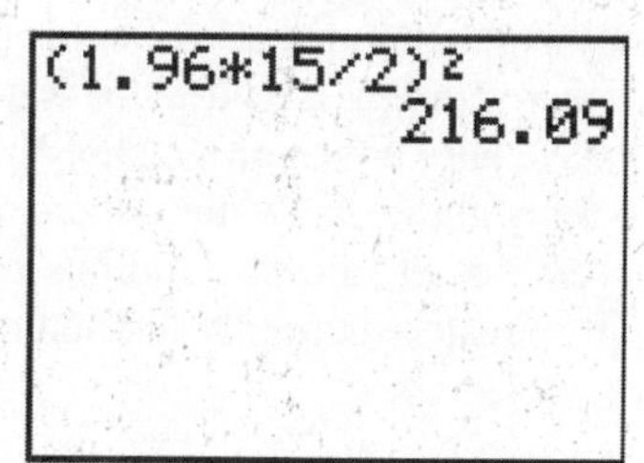

We have found before that for 95% confidence, $Z_{\alpha/2} = 1.96$. We also have $\sigma = 15$, and $E = 2$. Putting these together with the formula (be careful with parentheses!) tells us we must sample at least 217 statistics professors to get the desired margin of error (remember, always "round" up!)

ESTIMATING A POPULATION MEAN: σ NOT KNOWN

The formula for the confidence interval used to estimate the mean of a population when the standard deviation σ is not known is also built into the calculators and is called `TInterval` (Option 8 on the `STAT, TESTS` menu of TI-83/84 calculators, option 2 on the `Ints` menu on TI-89s.) This built-in function can be used with either raw data or summary statistics just like the `ZInterval` already discussed. The interval could also be calculated on the home screen in a similar manner as was shown for the other confidence intervals in this section.

EXAMPLE Confidence Interval for Birth Weights: In a study of the prenatal effects of cocaine use on infants, the following sample data were obtained for weights at birth: $n = 190, \bar{x} = 2700$ g, $s = 700$ g. The design of the study justifies the assumption that the sample can be treated as a simple random sample. Construct a 95% confidence interval for μ, the mean weight of all infants born to mothers who use cocaine. In this example we do not know σ, but we do have a value of $s = 700$ g which we can use as an estimate for σ.

TInterval with Summary Statistics

Press STAT ▶ ▶ 8 to get the input screen for a `TInterval`. Your first choice (just as with the `ZInterval`) is the type of Input you will use. For this problem you have summary statistics (not raw data), so you should highlight on "Stats" and press ENTER. Now fill in all the other information which was given in your problem as you see it on the screen.

```
TInterval
 Inpt:Data Stats
 x̄:2700
 Sx:700
 n:190
 C-Level:.95
 Calculate
```

Highlight the word `Calculate` and press ENTER for the results. We are 95% confident the mean weight of babies whose mothers use cocaine is between 2599 g and 2800 g.

Note This particular interval varies <u>very</u> little from what one would get using a `ZInterval`. This is because the sample size 190 is quite a large sample and the T distribution grows closer to the Z distribution as the sample size grows larger.

```
TInterval
 (2599.8,2800.2)
 x̄=2700
 Sx=700
 n=190
```

To retrieve the margin of error that was used to build the interval, we again must find the difference between the upper and lower endpoints and divide this difference by 2. The margin of error was 100.5.

TInterval with Raw Data

EXAMPLE Constructing a Confidence Interval: The data below are the ages of applicants who were unsuccessful in winning promotion (Barry and Boland, *American Statistician,* Vol. 58, No. 2)

34	37	37	41	42	43	44
45	45	45	46	48	49	53
53	54	54	55	56	57	60

Put the data into list `L1`. (Do this by typing it in or transferring from another source.) Because this is a small sample ($n < 30$), we should check to make sure it is approximately normal (no outliers or strong skewness). Looking at the data values, there do not appear to be any unreasonably large or small values. A normal plot of the data also shows the data are approximately normally distributed.

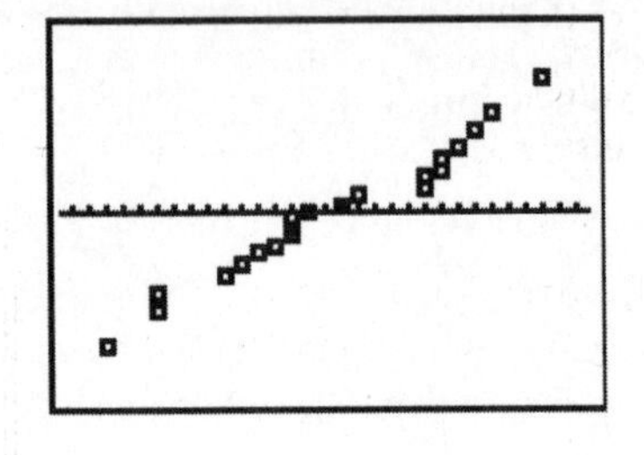

Press [STAT], arrow to TESTS and select option 8 (Ints and option 2 on an 89) to get a screen for the TInterval. Your first choice is the type of Input you will use. For this problem you have raw data, so you should move the cursor to highlight Data and press [ENTER] to move the highlight. Now fill in all the other information. This time you are not asked to specify σ because it is supposed to be unknown when using the TInterval. You must, of course, specify the list where you have stored the data.

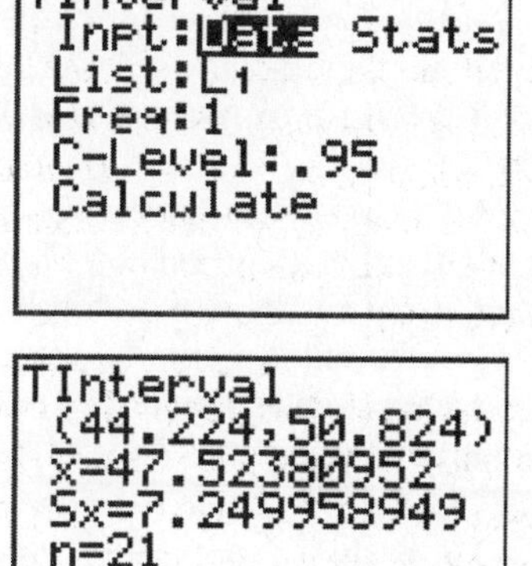

Highlight the word Calculate and press [ENTER]. Based on this sample, the average of those who were denied promotion is between 44.2 and 50.8 years old, with 95% confidence.

Home Screen Calculations for Mean Estimates

To calculate the margin of error for the above example on the home screen we would need to find the value of $t_{\alpha/2}$ for a 95% confidence level with (n-1) = 20 degrees of freedom. TI-89 and TI-84 calculators have a built-in inverse t function on the DISTR menu to find this value. If you are using a TI-83 you will either have to use a table or solve for a critical value as described below.

Solving for Critical Values from the T Distribution (TI-83)

Press [MATH] [0] to get the Solver from the MATH menu. You should see a screen with EQUATION SOLVER at the top. If not, use the [▲] key to cursor up. Next paste in the function tcdf(from the DISTR menu by pressing [2nd] [VARS] [5] . Fill in the rest of the values as at right. (We are asking the calculator to find the value of X from the t distribution with 20 degrees of freedom which has an area of .025 to its right.)

Press [ENTER] to see the screen at right. Type 2 as the first guess for the prompt X=.

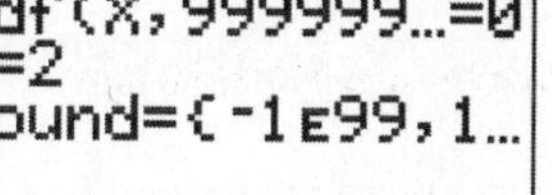

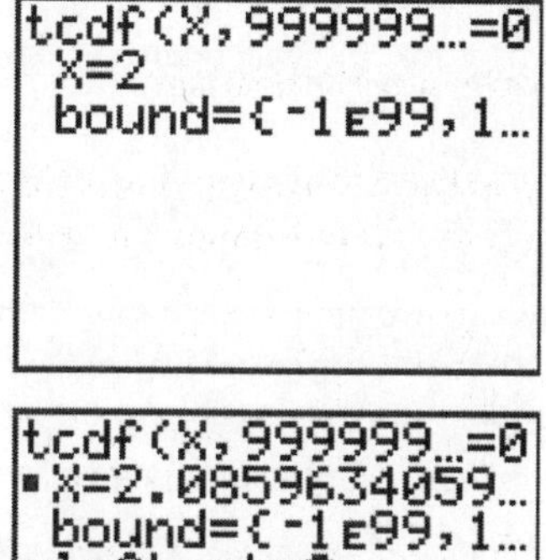

With the cursor flashing on the X=2 line, press [ALPHA] [ENTER] (SOLVE). Then you must wait for the calculator to find X (which is the $t_{\alpha/2}$ value you desire.) Be patient – it may take a while! Finally, you will see the answer. We find X = 2.08596.

Once you have found the value of $t_{\alpha/2}$, the home screen calculation of the margin of error and confidence interval are very straight-forward and proceed like that for a Zinterval as explained above.

Critical Values from the T Distribution – TI-84/89

Both of these calculators have built-in functions to obtain critical values from *t* distributions. On the TI-84, invT is option 4 on the [2nd] ❺ (DISTR) menu. On a TI-89, inverse distributions are option 2 on the [F5] (Distr), menu. Inverse t is option 2 from that submenu. In both cases, the parameters to enter are the area to the left of the point of interest, and the degrees of freedom (*n* - 1 for a single sample). The input dialog box for the *t*-value for a 95% confidence interval for our example above (*n* = 21, so df = 20) is at right. Notice, since t-distributions are symmetric around 0, we could also have used .025 as the area to the left, and then ignored the negative sign.

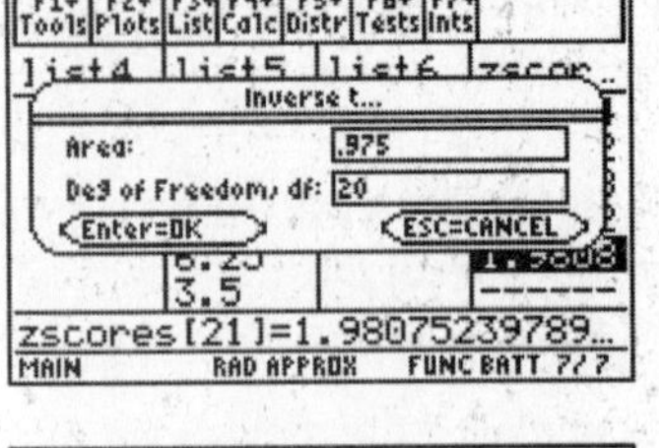

The calculator reports the critical value for this interval is 2.08596 as found above with the solver.

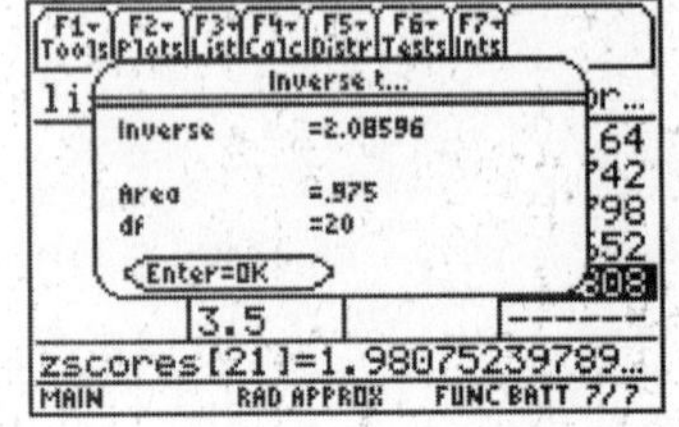

ESTIMATING A POPULATION VARIANCE

Methods for computing confidence intervals for a population variance (or standard deviation) are not built-in to TI calculators. These require critical values from a χ^2 (Chi-squared) distribution. Because the χ^2 distributions are not symmetric, these intervals are not of the usual *estimate* ± *ME*. Computation of these intervals is very sensitive to departures from the assumption the data comes from a normal distribution. This is a critical assumption to check – do a Normal quantile plot as described in Chapter 6.

EXAMPLE Penny Weights: Pennies are currently being minted with a standard deviation of 0.01654 g (based on the data from Data Set 14 in Appendix B). New equipment is being tested in an attempt to improve quality by reducing variation. A simple random sample of 10 pennies is obtained form the pennies minted by the new equipment. A normal quantile plot shows the weights came from a normally distributed population, and the standard deviation of the weights of the 10 pennies is 0.0125 g. Construct a 95% confidence interval estimate for σ, the standard deviation of the weights of the whole population of pennies made with the new equipment.

With degrees of freedom (10 – 1) = 9, we can calculate the two critical values from the chi-square distribution which are required in this confidence interval calculation. Using a TI-83 or -84, we can use the Solver as we did when finding critical *t*-values in the section above. The TI-89 has an inverse Chi-squared function built in, use it just as in finding critical *t*-values.

Since the χ^2 distributions are not symmetric, we need to find two critical values: χ^2_L with 0.025 area to the left, and χ^2_R with 0.975 area to the left. (The area between these points is our desired 95%.)

Finding χ^2 Critical Values TI-83/84 Procedure:

Press [MATH] then press the up arrow to find the Solver at option 0. Press [ENTER] to select the option. Press the up arrow to enter the equation to be solved.

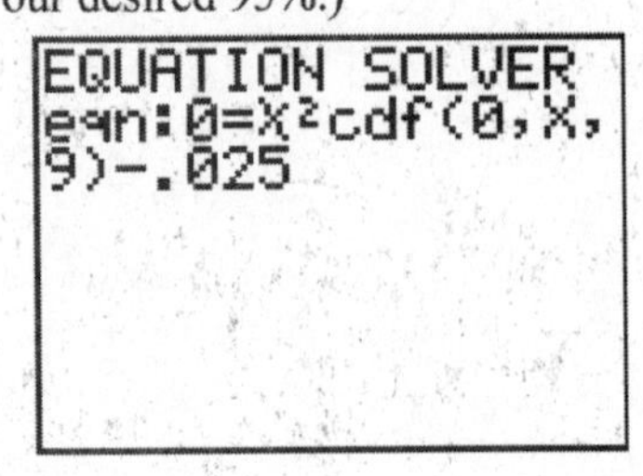

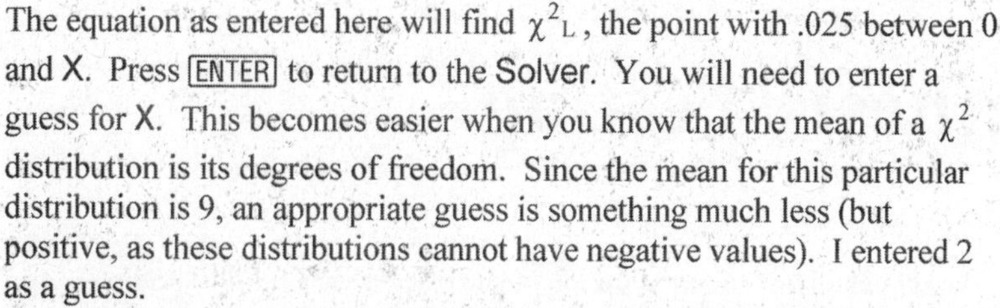

The equation as entered here will find χ^2_L, the point with .025 between 0 and X. Press [ENTER] to return to the Solver. You will need to enter a guess for X. This becomes easier when you know that the mean of a χ^2 distribution is its degrees of freedom. Since the mean for this particular distribution is 9, an appropriate guess is something much less (but positive, as these distributions cannot have negative values). I entered 2 as a guess.

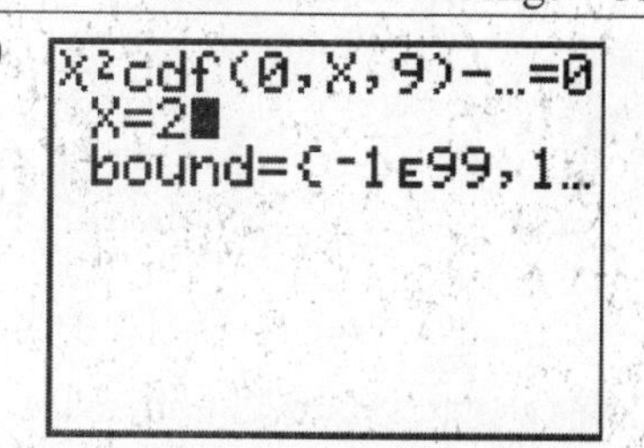

Press [ALPHA][ENTER] to solve for X. (It may take a little while.) χ^2_L is 2.7003895. To find χ^2_R, repeat the procedure, but change 0.025 in the equation to 0.975 and use a larger guess (bigger than 9 – try 15, for example). I found χ^2_R to be 19.022767798.

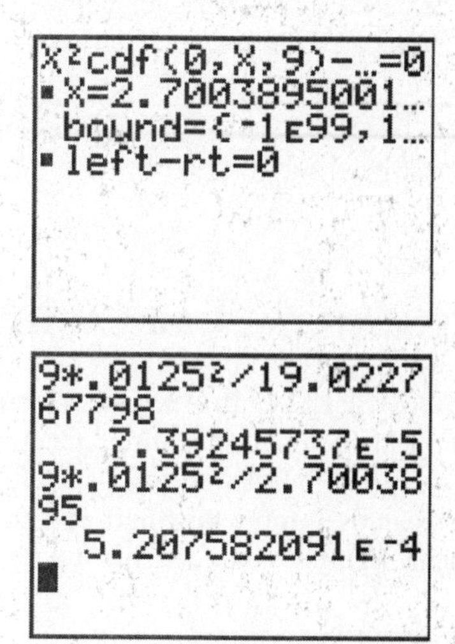

These results can be used to calculate the confidence interval limits using the formula $\frac{(n-1)s^2}{\chi^2_R} < \sigma^2 < \frac{(n-1)s^2}{\chi^2_L}$. Our values are very close to the value obtained in the text, but more accurate as the text used *approximated* critical values. The variance of the weights of the pennies from the new equipment is between 0.0000739 and 0.0005208. Taking the square root of both ends, gives a 95% confidence interval for the standard deviation of the weights of the pennies from the new equipment is 0.0085 to 0.0228 g.

WHAT CAN GO WRONG?

Assumptions not met

It is important to check that assumptions are met before conducting any inferential procedure. If for a proportion, are there at least 5 successes and 5 failures (so we can believe $\hat{p}$ is approximately normal)? In doing inference for a population mean, can we believe that $\bar{x}$ is approximately normal? You need a sample size of at least 30 for the Central Limit Theorem to apply. If you have a smaller sample, plot the data to check the assumption. This plot exhibits a skewed distribution. The inference methods we have discussed would not be appropriate.

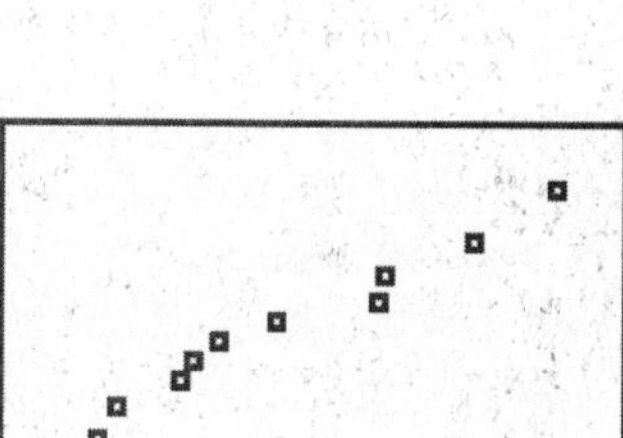

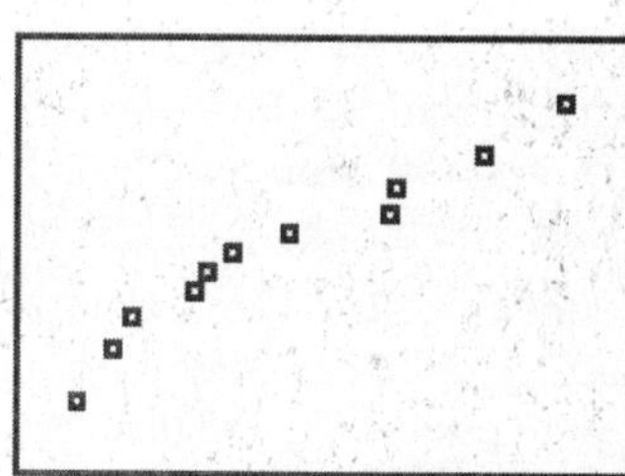

Bad Interpretations

Be careful when interpreting the meaning of any confidence interval. One must include the level of confidence, the parameter of interest (what the question is about), and units of measure. An interval for a mean, for example, is the mean of the entire population of interest – not the mean of the sample (we *know* what that was), or an individual value.

8 Hypothesis Testing

In this chapter, you will learn to use your TI calculator to assist you in performing a variety of hypothesis tests based on one sample. You will see how to use the tests built into the STAT TESTS menu as well as how to calculate test statistics on the home screen. You will be able to handle samples in the form of raw data as well as summary statistics. When performing a test you should always make sure that all assumptions are met. Refer to your main text for this material. There are many options available, so make sure you are following a routine which will allow you to include all steps required by your instructor in your write-up of a test.

TESTING A CLAIM ABOUT A PROPORTION

EXAMPLE Finding an Job Through Networking: A survey of 703 randomly selected workers showed that 61% got their jobs through networking. Based on these sample results, test the claim that most (more than half) of all Americans get their jobs through networking, that is $p > 0.5$. Although the observed 61% in the sample is more than half, is it enough bigger that we can really believe the population proportion is more than a half? We deduce that we are to test the following hypotheses: H_0: $p = 0.50$ and H_1: $p > 0.50$.

TI-83/84 Procedure

We will use the built-in test for a population proportion from the STAT TESTS menu. This is option 5:1-PropZTest. This test yields output which is perfectly-suited for the p-value method of testing a hypothesis.

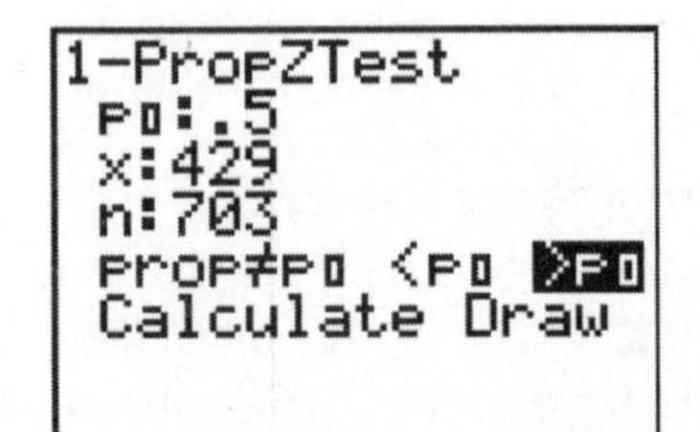

Press [STAT], arrow to TESTS, and select option [5]. Look at the inputs which are required. First is p_0, the value of the proportion *according to* H_0. This is 0.50 in our problem. Second is the value of x, the number sampled with the characteristic of interest. We know that $x = n\hat{p} =$ (703)(0.61) = 428.83 (round up to 429, since there cannot be fractions of a success). You can enter the multiplication on the input screen, press [ENTER] to perform it, and then use the up arrow to do the necessary rounding. Third is the value of n which is 703. Fourth, we must choose the direction of our H_1. This is >pø in our problem. Choices correspond to two-tailed, left-tailed and right-tailed tests. Use the left or right arrows to move the cursor to the correct choice, and press [ENTER] to move the highlight. Finally, we have the option of simply calculating our test results or obtaining a drawing of our results. Either choice will do. Fill in the required inputs.

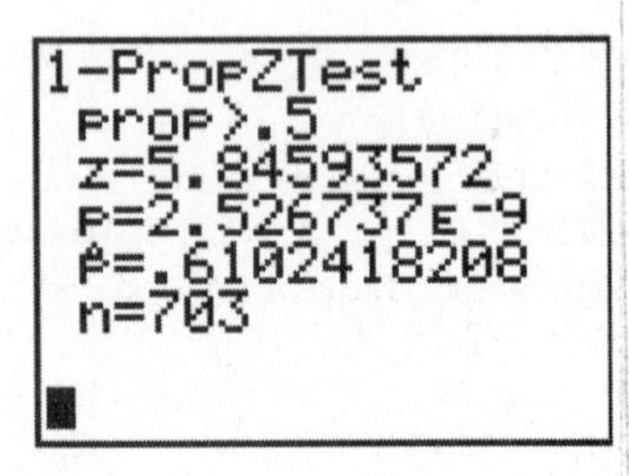

If you choose Calculate and then press [ENTER], you will see this screen. The computed *z*-statistic is 5.846. (Our observed 61% is more than 5 standard deviations above the claimed value of 50%. We know this is a rare event). The p-value for the test (the probability of finding our observed 61% (or something more extreme), if the value really was 50% is 2.5×10^{-9}, or 0.0000000025. This is an extremely small probability. This leads us to believe the true percentage of people who find jobs through networking is more than 50%.

If you choose `Draw` and then press [ENTER], you will see this screen. We are shown the z-statistic for the test and the p-value (to 4 significant digits). Notice the p-value (0.0000000025) is shown as 0. The area corresponding to the p-value of the test is shaded in the graph (like using ShadeNorm for normal calculations). Nothing is visibly shaded in this graph, due to the small p-value.

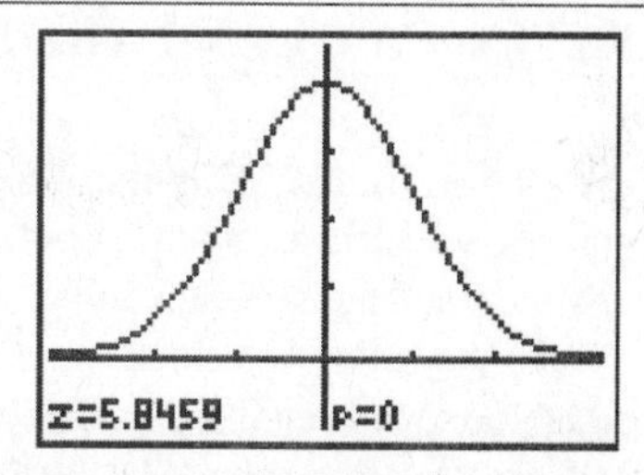

Our conclusion about networking for finding jobs? We will conclude that most people *do* find jobs through networking. Be sure to ask your instructor about the type of conclusion he or she might want in any situation.

TI-89 Procedure

Since the results of the survey were given as a percentage, you need to multiply to find the actual number of individuals who used networking to find their jobs. `x` = $n\hat{p}$ = (703)(0.61) = 428.83 (round up to 429). This cannot be done in the input dialog box for the test, but must be done on the home screen.

From the Statistics/List Editor, press [2nd][F1] (= [F6] `Tests`). Press [5] to select `1-PropZTest`. First is `p0`, the value of the proportion *according to H_0*. This is 0.50 in our problem. Second is the value of `x`, the number sampled with the characteristic of interest. We just computed that to be 429. Third is the value of `n` which is 703. Fourth, we must choose the direction of our H_1. Use the right arrow key to expand and select the proper direction. Lastly, you can decide to simply calculate the results, or display a normal curve with the area corresponding to the p-value shaded.

```
1-Proportion Z Test
p0:               .5
Successes, x:     429
n:                703
Alternate Hyp:    prop > p0
Results:          Calculate
Enter=OK          ESC=CANCEL
zscores[21]=1.98075239789...
MAIN      RAD APPROX      FUNC BATT
```

The screen at right is the output from selecting `Calculate`. The results are identical to those obtained with an 83 or 84.

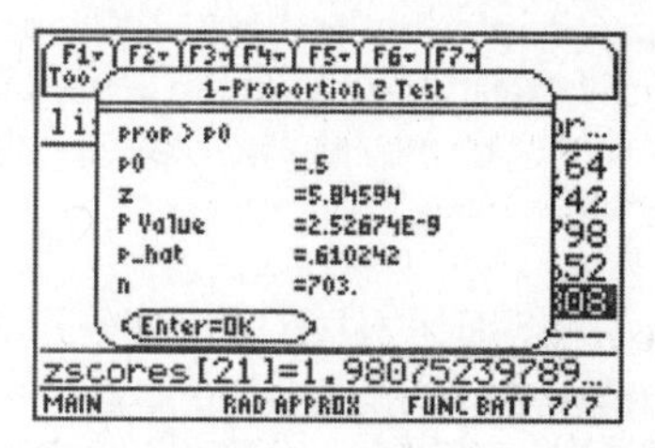

Home Screen Calculations

You can calculate the test statistic on the home screen using the formula $z = \dfrac{\hat{p}-p_0}{\sqrt{\dfrac{p_0 q_0}{n}}} = \dfrac{0.61-0.5}{\sqrt{\dfrac{(0.5)(0.5)}{703}}}$. We find our test statistic is $z = 5.833$. (The difference from those above is due to rounding). You can calculate the p-value of this test statistic by finding the area (probability) in the tail of the standard normal distribution corresponding to the direction of the alternate hypothesis (here, greater than the observed value) past the statistic as in Chapter 6.

```
(.61-.5)/√(.5*.5/703)
                5.833112377
normalcdf(5.833,E99
                2.730644583E-9
```

Also (for those using the Traditional Method) you can calculate the critical value associated with a particular significance level (in this case $\alpha = 0.05$) using the command `invNorm(.95`, since on a right-tailed test like this, our α area is on the right side of the curve, so 0.95 is the area to the *left* of the critical value.

Note: The Traditional Method would lead us to compare the critical value 1.645 and the test statistic 5.833. We find that the test statistic is greater and thus leads us to reject H_0 in favor of H_1 since this is a right-tailed (>) test.

Finally, refer to Chapter 7 for details if you wish to use the Confidence Interval Approach to testing.

TESTING A CLAIM ABOUT A MEAN, σ KNOWN

EXAMPLE M&M Weights: Data Set 13 in Appendix B includes weights of 13 red M&M candies randomly selected from a bag containing 465 M&Ms. The standard deviation of the weights of all of the M&Ms in the bag is $\sigma = 0.0565$ g. The sample weights (in grams) are listed below, and they have a mean of $\bar{x} = 0.8635$ g. The bag states that the net weight of the contents is 396.9 g, so the M&Ms must have a mean weight that is at least 396.9/465 = 0.8535 g in order to provide the amount claimed. Use the sample data with a 0.05 significance level to test the claim of a production manager that the M&Ms have a mean that is actually larger than 0.8535 g, so consumers are being given more than the amount indicated on the label.

0.751	0.841	0.856	0.799	0.966	0.859	0.857	0.942	0.873	0.809	0.890	0.878	0.905

We will use the calculator's built-in test for a population mean when σ is known. This is option 1 from the STAT TESTS menu. It is the ZTest. As in the previous example, this test yields output which is perfectly-suited for the p-value method of testing a hypothesis. This procedure is analogous, whether you are using a TI-83/84 or a TI-89 calculator.

As before when you worked with confidence intervals for a mean, you have two options for input. You can input summary statistics or you can use the raw data as input. If you wish to use raw data, you must have it stored in a list. For our example below, we have stored the 11 weights in list L1.

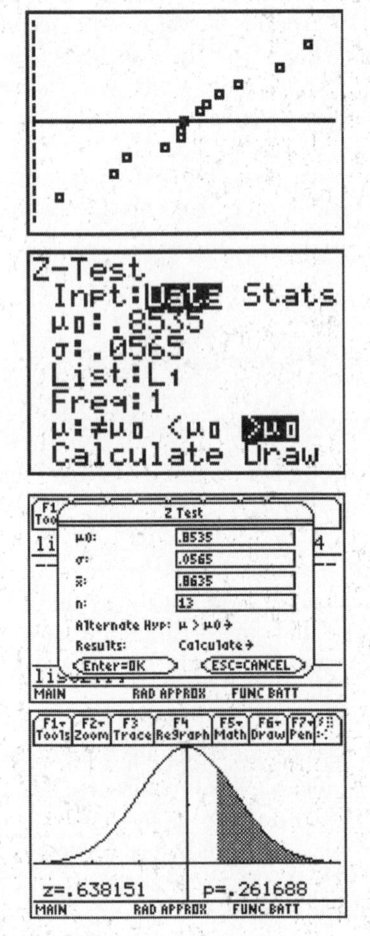

Requirements Check: This is a small sample (n = 13), so even though we know σ, we must check to determine whether our sample is from an (at least approximately) normal population. I have created a normal quantile plot of the data (see Chapter 6 for details). The plot at right is fairly linear, so the normal assumption is satisfied, and we can proceed with a test.

Press STAT ▶ ▶ 1 to get the input screen (2nd F1 = F6 (Tests) followed by ENTER on an 89). If you choose to input Data and press ENTER your screen will be like this one.

In this TI-89 dialog box, we have chosen to input summary statistics. Note that we still had to specify the *hypothesized mean* $\mu_0 = 0.8535$ g. We also had to specify the population variance $\sigma = 0.0535$ g because it is supposed to be known when using the Z-test.

Fill in the rest of the information requested. Note this is a one-tailed test (the claim is that the bags contain more weight than stated on the label), so we choose $> \mu_0$ as our alternative hypothesis.

When all is filled in, choose either Calculate or Draw and press ENTER. The results of "Draw" (this author's personal favorite option) are shown. Our computed *z*-statistic is $z = 0.638$, with a p-value of 0.2617. This large probability tells us that there is a good chance of obtaining our observed sample mean ($\bar{x} = 0.8635$ g) if the population mean is 0.8535 g. We fail to reject the null hypothesis; we are not convinced the bags of candy weigh more than they are labeled. You may have slightly different values (due to rounding on the summary statistics) but the conclusion is still the same.

Home Screen Calculation

We have reproduced the calculations for the home screen calculations for the test statistic $z = \dfrac{\bar{x} - \mu_0}{\frac{\sigma}{\sqrt{n}}}$ and it's p-value at right. The author does not recommend this approach, as misplacing parentheses and failing to double the p-value for a two-sided test are potentially fatal errors – I recommend the automated approach.

```
(.8635-.8535)/(.
0565/√(13))
             .6381506682
normalcdf(.63805
,E99
             .2617205017
```

TESTING A CLAIM ABOUT A MEAN, σ NOT KNOWN

EXAMPLE M&M Weights: Data Set 13 in Appendix B includes weights of 13 red M&M candies randomly selected from a bag containing 465 M&Ms. The standard deviation of the weights of the sample is $s = 0.0576$ g. The sample weights (in grams) are listed below, and they have a mean of $\bar{x} = 0.8635$ g. The bag states that the net weight of the contents is 396.9 g, so the M&Ms must have a mean weight that is at least 396.9/465 = 0.8535 g in order to provide the amount claimed. Use the sample data with a 0.05 significance level to test the claim of a production manager that the M&Ms have a mean that is actually larger than 0.8535 g, so consumers are being given more than the amount indicated on the label.

0.751	0.841	0.856	0.799	0.966	0.859	0.857	0.942	0.873	0.809	0.890	0.878	0.905

We will use the calculators' built-in test for a population mean when σ is not known. This is option 2 from the STAT TESTS menu. It is the T-Test. As in the previous examples, this test yields output which is perfectly-suited for the p-value method of testing a hypothesis.

As in the preceding example, you have two options for input. You can input summary statistics or you can use the raw data as input. If you wish to use raw data, you must have it stored in a list. For this example, we have summary statistics and raw data (which, as we saw above, satisfy the assumption that the sample comes from a normal population), so we can use either method.

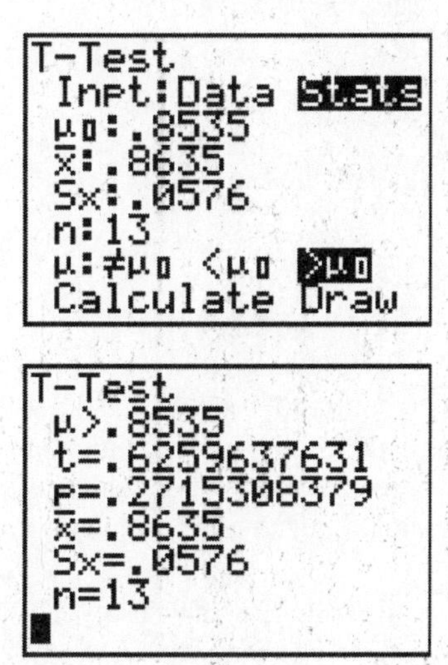

Press [STAT] [▸] [▸] [2] ([2nd][F1][2] from the Statistics/List Editor on an 89). Here, I have chosen Stats as the input method. To move the highlight, move the blinking cursor to your choice and press [ENTER].

Specify the *hypothesized mean* $\mu_0 = .8535$. This time we do not have to specify the population standard deviation σ (but we do have to input *s* because we have chosen the Stats input method. Note this is a right-tailed test, so we choose $>\mu_0$ as our alternative hypothesis.

When all is filled in, choose either Calculate or Draw and press [ENTER]. The results of Calculate are shown at right. Our test statistic is $t = .626$, with a p-value of 0.2715. We still fail to reject the null hypothesis, and conclude that the weights of M&Ms bags are not more than they are labeled.

To use the traditional (rejection region) approach, we need a t-critical value to compare our calculated statistic against. The critical value can be obtained from tables or by using equation solver as described in Chapter 7 (TI-83) or the Inverse T function (TI-84 and 89). In the screen at right, I have done the computation using a TI-84. The parameters are the area to the left of the desired point (since our test at $\alpha = 0.05$ was right tailed, all our α area is on the right, so 0.95 to the left) and with $n = 13$ pieces of data, we have $n - 1 = 12$ degrees of freedom. Since our computed statistic ($t = 0.626$) is less than the critical value, we fail to reject the null hypothesis.

TESTING A CLAIM ABOUT A STANDARD DEVIATION OR VARIANCE

Just as with confidence intervals for a variance or standard deviation, there is no built-in function on TI-calculators to conduct a hypothesis test for a variance or standard deviation. The Chi-square cdf function on all calculators can help find exact p-values, however, and the Equation Solver (on TI-83/84s) and Inverse Chi-Square on TI-89s can help find critical values. As with confidence intervals for a variance, this procedure is very sensitive to any departures from the normal distribution assumption – be sure to check plots of your data to be sure this is satisfied!

EXAMPLE Quality Control: The industrial world shares this common goal: Improve quality by reducing variability. Quality-control engineers want to ensure that a product has an acceptable mean, but they also want to produce items of consistent quality so that there will be few defects. The Newport Bottling Company has been manufacturing cans of cola with amounts having a standard deviation of 0.051 oz. A new bottling machine is tested, and a simple random sample of 24 cans results in the amounts (in ounces) listed below. (Those 24 amounts have a standard deviation of $s = 0.039$ oz.) Use a 0.05 significance level to test the claim that cans of cola from the new machine have amounts with a standard deviation that is less than 0.051 oz.

11.98	11.98	11.99	11.98	11.90	12.02	11.99	11.93	12.02	12.02	12.02	11.98
12.01	12.00	11.99	11.95	11.95	11.96	11.96	12.02	11.99	12.07	11.93	12.05

Create a plot to check the normal distribution assumption. The normal quantile plot at right is linear, so there is no departure from the normal distribution assumption. We may proceed with a test.

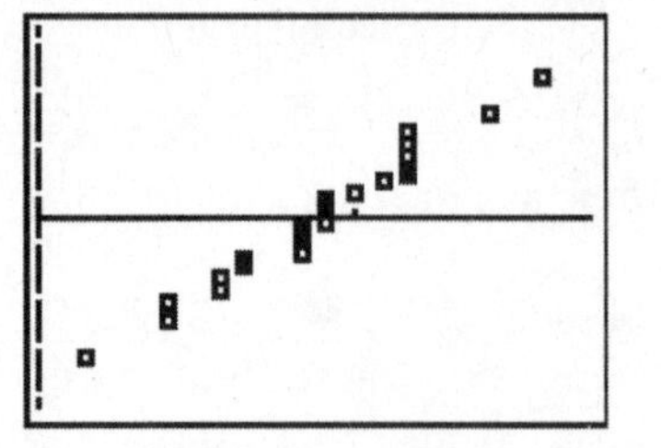

Calculate the test statistic from the following equation:

$\chi^2 = (n-1)\dfrac{s^2}{\sigma^2} = (24-1)\dfrac{.039^2}{.051^2} = 13.4498$ as seen in the top of the screen at right.

```
(24-1)*.039²/.05
1²
         13.44982699
X²cdf(0,13.4498,
23)
        .0584478992
```

The bottom of this screen shows the calculation of the p-value of the test statistic using the χ^2cdf function located in the DISTR menu. Note we wanted area *below* our calculated statistic because the claim is that the standard deviation is now *lower*.

Finally, for those using the traditional method, you can find the critical values using the Equation Solver (TI-83/84) or Inverse Chi-Square as detailed in Chapter 7.

The p-value is 0.05845 which is not less than the significance level of the test. The test statistic (13.450) is more than the left critical value of the test (13.0905), so either way you look at it, the sample has provided evidence that the standard deviation of the new bottling machine is not less than 0.051.

WHAT CAN GO WRONG?

Err: Domain?
This error stems from one of two types of problems. Either a proportion was entered in a `1-PropZTest` which was not in decimal form or the numbers of trials and/or successes was not an integer. Go back to the input screen and correct the problem.

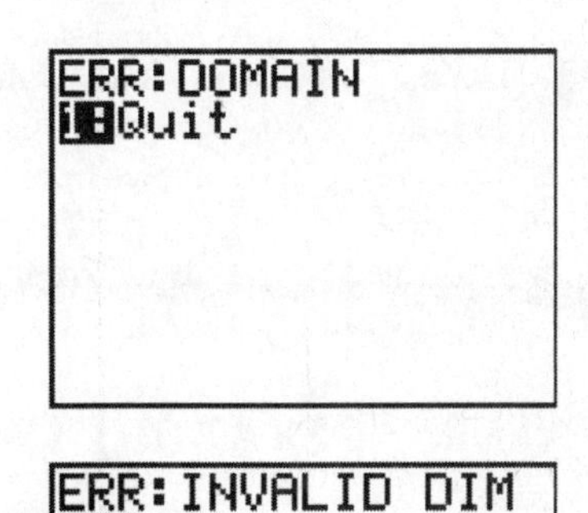

Err:Invalid Dim?
This can be caused by selecting the `DRAW` option if another Statistics plot is turned on. Either go to the `STAT PLOT` menu ([2nd][Y=]) and turn off the plot or redo the test selecting `CALCULATE`.

ERR:INVALID DIM
1:Quit

Bad Decisions.
It is important to remember that we have calculated a test statistic *assuming the null hypothesis is true*. The p-value of the test is the probability of our sample result (or something more extreme) under that assumption. If our sample result is unlikely to have happened, we reject the null hypothesis in favor of the alternate. That is, reject H_0 for ***small*** p-values.

Bad Conclusions.
If the p-value is large, the null hypothesis is not rejected; but this does *not* mean it is true – we simply haven't gotten enough evidence to show it's wrong. Be careful when writing conclusions to make them agree with the decision – remember, the null hypothesis says the claimed value is *true*. If H_0 is rejected, this means you no longer believe this value, but rather believe the true value (mean or proportion) is different (or higher, or less).

9 Inferences from Two Samples

In Chapters 7 and 8 you found confidence intervals and tested hypotheses for data sets that involved only one sample from one population. In this chapter you will learn how to extend the same concepts for use on data sets that involve two samples from two populations. As in Chapters 7 and 8 you will use functions from the STAT TESTS menu. The presentation in this chapter assumes a familiarity with the materials presented in Chapters 7 and 8. As always, if there are significant differences, the TI-89 procedure is explained immediately after that for the TI-83 and -84.

INFERENCES ABOUT TWO PROPORTIONS

EXAMPLE Is Surgery Better than Splinting?: The Chapter Problem includes results from a clinical trial in which patients were treated for carpal tunnel syndrome, and Table 9-1 summarized the results. They are repeated here for convenience. We want to test at a 0.05 significance level the claim that the success rate is better with surgery than with splinting.

	Surgery	**Splint**
Success one year later	67	60
Total number treated	73	83
Success Rate	92%	72%

Hypothesis Test: 2-PropZTest

Press [STAT] [►] [►] [6] to get to the STAT TESTS menu and choose option 6, the 2-PropZTest (on a TI-89, from the Statistics/List Editor press [2nd] [F1] [6]). The first four input values come directly from the table. Next we choose the direction of our test hypothesis. We are asked to test if the proportion of successful cures with surgery is *greater than* the proportion of successful cures with splinting. Since we represented the surgery patients with x1 and n1, it would follow that our test hypothesis is p1>p2. Thus you should fill in the screen as at right.

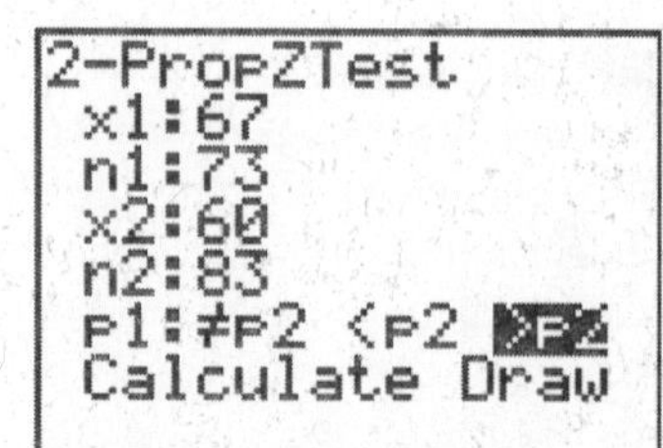

Highlight the word Calculate and press [ENTER] for the results. The test statistic is $z = 3.123$. Note the p-value is 8.96×10^{-4} or 0.000896. We will reject the null hypothesis and conclude that surgery is indeed better than splinting to cure carpal tunnel syndrome.

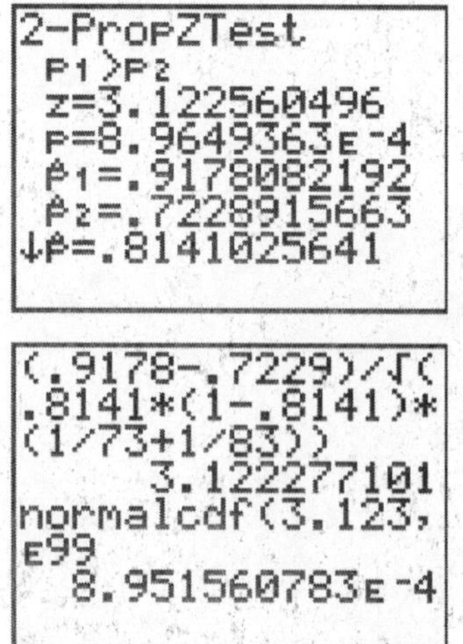

If one really wanted to, one could also do the calculation of the test statistic on the home screen. This first involves the calculation of $\bar{p} = \dfrac{x_1 + x_2}{n_1 + n_2} = \dfrac{67+60}{73+83} = .8141$. This was shown as $\hat{p}$ at the bottom or our results screen. Then we find the test statistic

$$Z = \frac{p1 - p2}{\sqrt{\bar{p}(1-\bar{p})\left(\dfrac{1}{n1} + \dfrac{1}{n2}\right)}}$$. Again, we obtain Z = 3.123.

Since the p-value is less than the significance level of 0.05, we find that our data has provided us with significant evidence that the proportion of successful cures of carpal tunnel syndrome is greater with surgery than with splinting.

Confidence Interval: 2-PropZInterval

EXAMPLE Is Surgery Better than Splinting?: Use the sample data in Table 9-1 to construct a 90% confidence interval for the difference in the two population proportions. (A 90% confidence interval is similar to a 95% 1-tailed hypothesis test.)

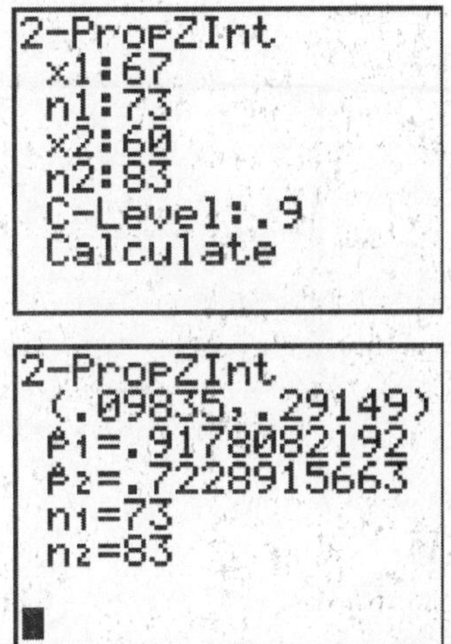

Press [STAT] [▸] [▸] for the `STAT TESTS` menu. Choose Option `B` which is the `2-PropZInt`. (This is option 6 on the `Ints` menu on a TI-89.) Fill in the screen as shown.

Highlight `Calculate` and press [ENTER] for the interval. We obtain the results (.09835, .29149). To retrieve the margin of error we can simply find the difference in the endpoints and divide by 2.
$E = (.29149 - .09835)/2 = .097$.

We find that somewhere between 9.8% and 29.2% more patients would be cured with surgery over splinting, at 90% confidence. Be sure to notice that both ends of this interval are positive. If there were no difference between the two treatments, the interval would include 0, which is not the case here.

INFERENCES ABOUT TWO MEANS: INDEPENDENT SAMPLES

When you have a data set which is comprised of values from two independent samples of two populations it can at first seem difficult choosing how to begin. For the problems in your Triola text it is assumed that you do not know the values of the population variances σ_1^2 and σ_2^2. (In real life, this is almost always the case.) If you, in some future setting, know the values of the population variances you can use the `2SampZTest` and the `2SampZInt` procedures (options 3 and 9 on the TI-83/84 `STAT TESTS` menu, option 3 on the TI-89 `Tests` and `Ints` menus). With the idea of keeping things simple, we will not go into more detail or show examples on these procedures in this companion.

Hypothesis Testing: 2-SampTTest (Assuming $\sigma_1 \neq \sigma_2$, σ_1 and σ_2 not known)

EXAMPLE Discrimination Based on Age: The revenue commissioners in Ireland conducted a contest for promotion. The ages of unsuccessful and successful applicants are given below (based on data from Barry and Boland, *The American Statistician*, Vol. 58, No. 2). Some of the applicants who were unsuccessful in getting the promotion charged that the competition involved discrimination based on age. If we treat the data as samples from larger populations and use a 0.05 significance level to test the claim that the unsuccessful applicants are from a population with a *greater* mean age than the mean age of unsuccessful applicants, does there appear to be discrimination based on age?

Ages of Unsuccessful Applicants									
34	37	37	38	41	42	43	44	44	45
45	45	46	48	49	53	53	54	54	55
56	57	60							

Ages of Successful Applicants									
27	33	36	37	38	38	39	42	42	43
43	44	44	44	45	45	45	45	46	48
47	47	48	48	49	49	51	51	52	54

I have entered the ages of the unsuccessful applicants into list L1 and the ages of the successful applicants into list L2. The first step with these small samples is to determine whether they each come from approximately normal populations. Create a normal quantile plot (see Chapter 6 for details) for each set of data.

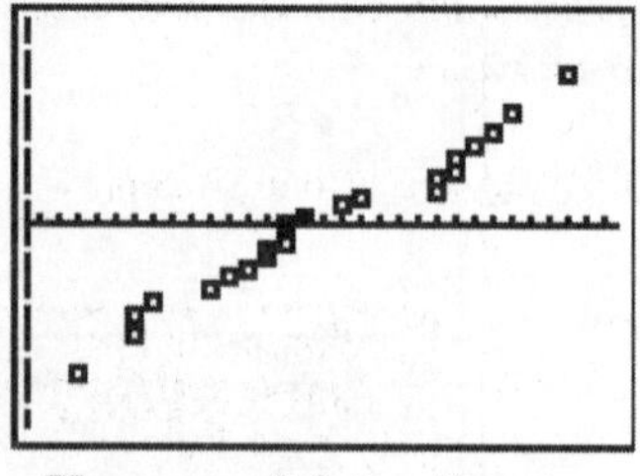

Unsuccessful Applicants

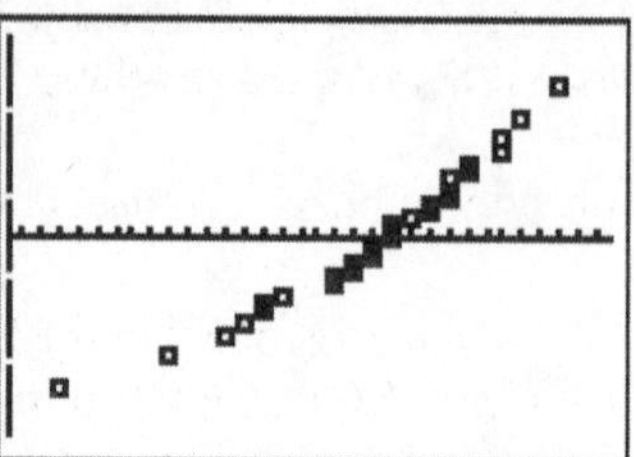

Successful Applicants

Both plots are reasonable close to straight lines, so we may proceed.

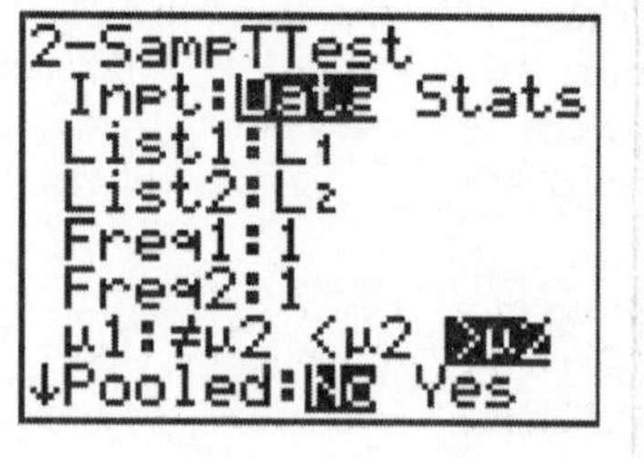

Press [STAT] [▸] [▸] [4] to choose Option 4:2-SampTTest from the STAT TESTS menu. (On a TI-89, this is also option 4). Since we have sets of raw data, move the cursor to highlight Data and press [ENTER] to move the highlight. (Input on the TI-89 is similar.)
Note: If you had only been given summary statistics (as is the case in some textbook problems), you would highlight Stats and press [ENTER]. Then you would be prompted to fill in all of the statistics for both data sets.

Match the inputs you see. Note we are telling the calculator which lists contain our two data sets. Near the bottom of the screen we choose the direction of the test hypothesis. This is a *one-tailed* test because the claim is that unsuccessful applicants were older. Also in the very bottom of the screen, we have chosen the answer "No" to the question of whether or not we will *pool* our standard deviations. (For homework exercises 9-28 in your text, you are advised to not assume the standard deviations are equal. This equates to not pooling.).

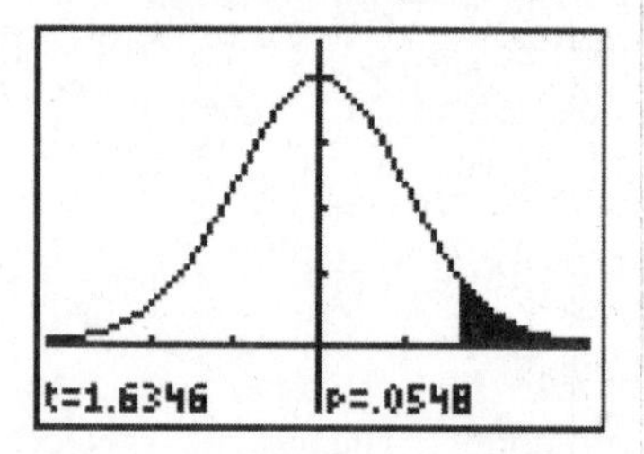

Press the [▾] key to see the last option. As usual for tests, you can choose either Calculate or Draw for displaying the results. We chose "Draw" and pressed [ENTER]. You can see the test statistic in the lower left corner is $t = 1.6346$. The p-value of 0.0548 is displayed in the lower right corner. It follows that we do not have sufficient evidence at the 0.05 significance level to conclude that the mean age of unsuccessful applicants was older than the mean age of successful applicants.

2-Sample T Test
μ1 > μ2
t =1.6346
P Value =.054815
df =41.8685
x̄1 =46.9565
x̄2 =43.9333
Sx1 =7.22041
Sx2 =5.88354
Enter=OK
MAIN RAD APPROX FUNC BATT 2/6

Your text suggests using the conservative approach to p-values for this situation. TI calculators (and most software) use a complicated formula for the degrees of freedom for these tests. As we can see in the screen at right (having chosen the Calculate option), the degrees of freedom (41.8685) are not even an integer. To find a p-value using the conservative approach (the smaller of $n_1 - 1$ and $n_2 - 1$ which would be 22 in this example) we can use the Tcdf function from the DISTR menu.

The results are displayed at right. This p-value does not change the conclusion; however there are cases in which this might occur. *Be sure to double the p-value obtained in this manner if you are working with a two-tailed alternate hypothesis.*

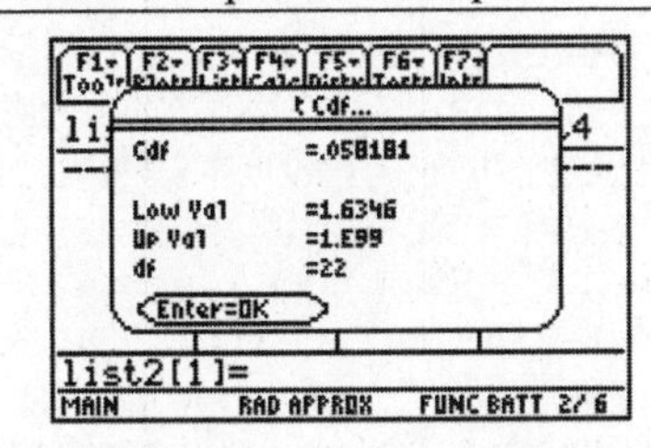

Confidence Interval: 2-SampTInterval (Assuming $\sigma_1 \neq \sigma_2$)

EXAMPLE Confidence Interval for Ages of Applicants for Promotion: Again using the data sets from on age of applicants for promotion, construct a 90% confidence interval estimate of the difference between the mean ages of those not promoted and those successful promotion candidates.

The two data sets are stored in lists L1(unsuccessful applicants) and L2 (successful applicants).

Press STAT ▸ ▸ 0 to choose Option 0 in the Stat Tests menu which is the 2-SampTInt. You have sets of raw data, so highlight Data and press ENTER to move the highlight, if needed.
Note: If you had only been given summary statistics (as is the case in some textbook problems), you would highlight Stats and press ENTER. Then you would be prompted to fill in all of the statistics for both data sets. Fill in the rest of the information to duplicate what you see in the screen. Note the near the bottom we indicate that our desired confidence level is 90%. Again, we answer No to the pooling question.

Press the ▾ key to highlight Calculate and press ENTER. You can see the confidence interval is (-.0878, 6.1341). We are 90% confident that the difference in the mean ages is between -.0878 and 6.1341 years. Since 0 is included in the interval, there is not a significant difference in mean age between those not promoted and those promoted.

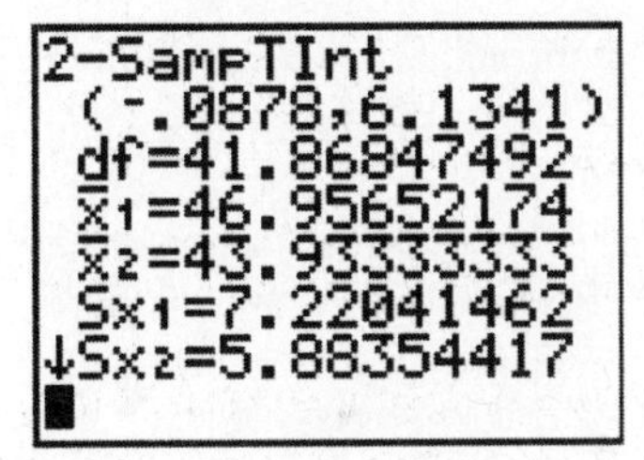

Home Screen Calculations – Conservative Approach.

The confidence interval computed with the built-in function uses the computed degrees of freedom. If you want to use the conservative approach, you will have to compute the interval "manually." You will first need a critical t-value – one that puts 5% (for 90% confidence, $\alpha = .10$, so $\alpha/2 = .05$) in each tail of the distribution. The smaller sample was for those not promoted, $n = 23$, so df = 22. If you have a TI-84 or TI-89, you can use the built-in invT function to find the critical value. If you have a TI-83, either use a table or use the Equation Solver as detailed in Chapter 7.

We first found the difference in the two sample means (46.957 – 43.933 = 3.024, using the means in the screen above). The endpoints of the interval are then calculated using the formula $\bar{x} \pm t_{\alpha/2}\sqrt{\frac{s_1^2}{n_1} + \frac{s_2^2}{n_2}}$.

From the screen at right, we see the conservative 90% confidence interval for the difference in mean age is (-.15, 6.20). As always, slight differences are due to rounding.

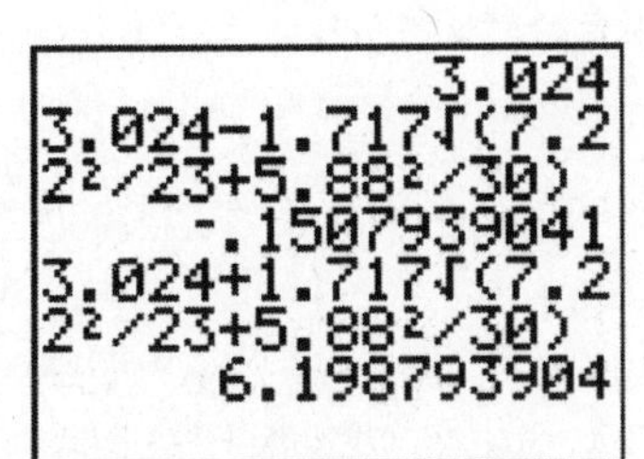

Hypothesis Testing and Confidence Intervals: (Assuming $\sigma_1 = \sigma_2$)

In problems 29 - 32 in section 9.3 of your text, you are told to assume the two independent samples came from normal populations with equal standard deviations ($\sigma_1 = \sigma_2$). In order to answer questions such as these, the only modification you should make to the above instructions is to answer `Yes` to the question of whether to pool or not. Your TI calculator will pool the two sample standard deviations and use the appropriate formulas for hypothesis test statistics and confidence intervals based on your instructions. In the end results, you will see the pooled standard deviation. It will be called Sxp.

INFERENCES ABOUT TWO MEANS: MATCHED PAIRS

EXAMPLE Are Forecast Temperatures Accurate?: The table below consists of 5 actual low temperatures and the corresponding low temperatures that were predicted five days earlier. The data consists of matched pairs, because each pair of values represents the same day. The forecast temperatures appear to be very different from the actual temperatures, but is there sufficient evidence to conclude that the mean difference is not zero? Use a 0.05 significance level to test the claim that there is a *difference* between the actual low temperatures and the low temperatures that were forecast five days earlier.

Actual Low	54	54	55	60	64
Low Forecast 5 days earlier	56	57	59	56	64
Difference (Actual – Predicted)	-2	-3	-4	4	0

We see that the above problem leads to the following hypotheses: H_0: $\mu_d = 0$ versus H_1: $\mu_d \neq 0$.

Put the actual low temperatures in `L1` and the predicted low temperatures in `L2`. Store the differences (`L1-L2`) in `L3`. Do this by highlighting `L3` in the Statistics Editor and then typing `L1-L2` and pressing [ENTER]. The results should be the same as those seen in the bottom row of the above table. Now you are ready to treat the differences as a single data set on which you can perform the desired test. The procedure for obtaining the differences is the same on the TI-89.

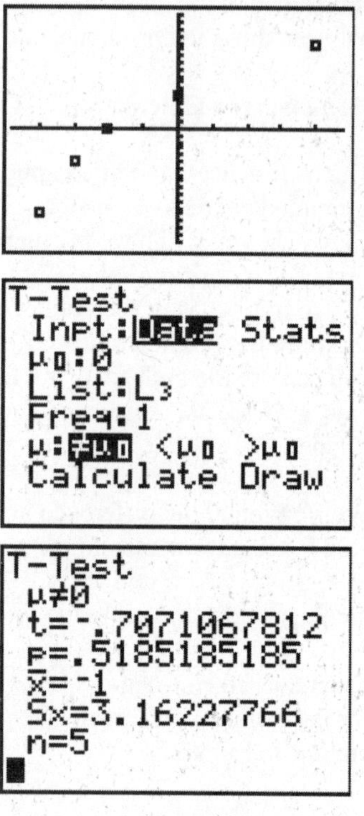

This is a small sample again. We (as always) need to check whether or not the assumption that the data came from a normal population is satisfied. A normal quantile plot is fairly straight, so we may continue. (Actually, this plot does show some curvature, but with the very small sample size, this is OK, as long as there are no clear outliers, and that is the case here.)

Press [STAT], arrow to `TESTS`, and select option `2:T-Test`. Set up the input screen to match mine. Note that we are telling the calculator that our data (the differences) is in `L3`. We are also specifying that the hypothesized mean μ_0 is 0 and that this is a two-tailed test (we want to know if the mean difference is *not* 0).

Highlight `Calculate` and press [ENTER] for the results. Note that the test statistic is $t = {}^{-}.707$ and the p-value is 0.5185. This is much larger than the significance level of 0.05, so our data was not strong enough evidence that there was an actual difference in the means. Also note that the values $\bar{x}$ and Sx are in this problem synonymous with $\bar{d}$ and Sd.

The test statistic can be calculated on the home screen by t = $\bar{d}$ - μ_0)/(Sd/ $\sqrt{n}$) = ‾13.2 – 0)/(10.686/√(5)) = ‾2.762. This, of course, would require your knowing the values of $\bar{d}$ and Sd. These could have been found using `1-Var Stats` from the `STAT CALC` menu for the differences in list `L3`.

Confidence Interval for μ_d

EXAMPLE Are Forecast Temperatures Accurate?: Use the same sample matched pairs to construct a 95% confidence interval estimate of μ_d which is the mean of the differences between actual low temperatures and five day forecast low temperatures.

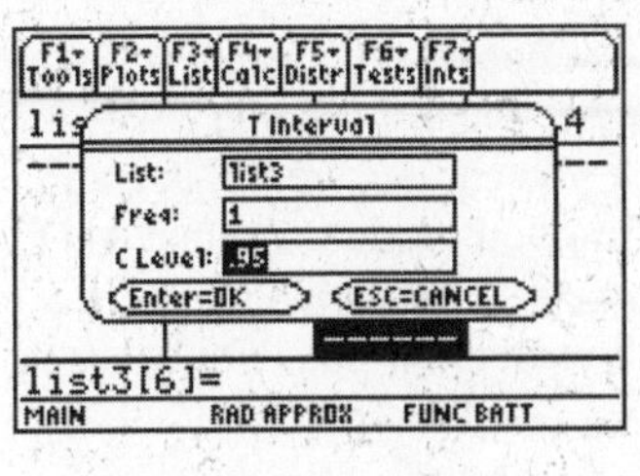

Press STAT, arrow to `TESTS` and choose Option 8 which is the `TInterval`. My input screen for the TI-89 is at right. The input for a TI-83/84 is analogous.

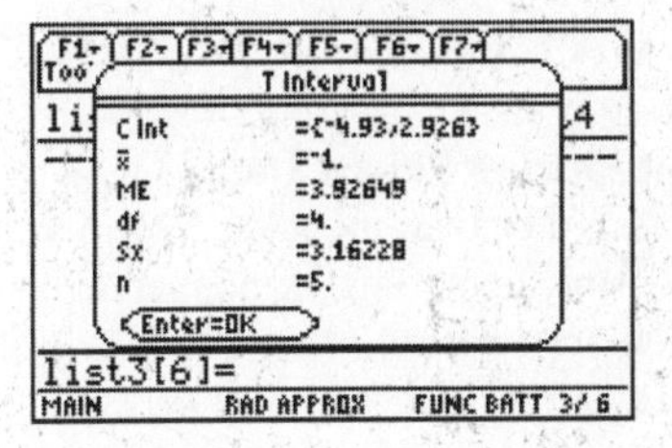

Press ENTER (after highlighting `Calculate` on an 83/84). The 95% confidence interval is given as (‾4.93, 2.926). Rounding, we can say we are 95% confidence that the mean difference in the actual and predicted low temperatures is between ‾4.9 and 2.9 degrees. This implies that the mean difference could be 0 (because 0 is contained in the interval). Again, our sample has not given sufficient evidence that there is a difference in the means.

The Home screen calculation of the margin of error of the above confidence interval would involve the formula $E = t_{\alpha/2} \frac{s_d}{\sqrt{n}} = 2.776 * 3.16/\sqrt{5} = 3.92$. The value 2.776 is from the t-table with (5-1) = 4 degrees of freedom.

COMPARING VARIANCES IN TWO SAMPLES

The following procedure is used to test if there is a difference in the variances of two populations based on two independent samples drawn from the populations. This test is sometimes used to decide whether or not to pool the standard deviations when comparing two population means. If this test determines the variances (and thus the standard deviations) are different then one would not want to pool the standard deviations. As always when working with variances, one must take extra care to ensure you are working with data from normal populations – this is a *very* critical assumption for this test.

EXAMPLE Coke Versus Pepsi: Data Set 12 in Appendix B includes the weights (in pounds) of samples of regular Coke and regular Pepsi. The sample statistics are summarized in the accompanying table. Use a 0.05 significance level to test the claim that the weights of regular Coke and regular Pepsi have the same standard deviation.

	Regular Coke	Regular Pepsi
n	36	36
$\bar{x}$	0.81682	0.82410
s	0.007507	0.005701

The above leads us to test the following hypotheses: H_0: $\sigma_1 = \sigma_2$ H_1: $\sigma_1 \neq \sigma_2$

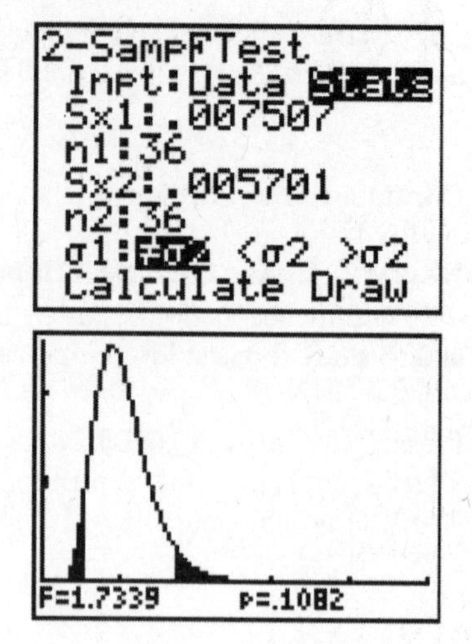

From the STAT TESTS menu, choose the option 2-SampFTest. The option number varies with the model of the calculator, but in all cases it is near the end of this menu. It is easiest to use the up arrow to locate it. Duplicate my input screen. (This time we are using the summary statistics instead of the raw data.)
NOTE: It is customary when using this test to use the larger standard deviation as the first sample.

If you choose to highlight Draw and press [ENTER], you will see the screen at right. The test statistic is F = 1.7339 and the p-value is 0.1082. One thing you can see from this graph is a major difference between F distributions and the normal and t distributions – F distributions are not symmetric. We conclude there is insufficient evidence to decide there is a difference in the standard deviations of the two kinds of cola.

WHAT CAN GO WRONG?

Not Pairing Paired Data
This is a critical mistake. One needs to think carefully if there is some natural pairing of data that might (possibly) come from independent samples. Clearly, if the samples sizes are not the same, the data cannot have been paired. If one fails to pair paired data, wrong conclusions will usually be made, due to overwhelming variability between the subjects.

Bad Conclusions
The biggest thing to guard against is bad conclusions. Think about the data and what they show. Do not let conclusions contradict a decision to reject (this means we believe the alternate is true) or not reject (this means we have failed to show the null is wrong) a null hypothesis. Also, one must be careful in keeping track of which sample was used as "group1" and which was "group2" in computing the test (and construction the alternate hypothesis).

Other than that, there is not much that hasn't already been discussed – trying to subtract lists of differing length will give a dimension mismatch error. Having more plots "turned on" than are needed can also cause errors.

10 Correlation and Regression

In this chapter we will see how TI calculators can help with simple linear regression and correlation. The calculator has a few built-in functions to aid in this work, but the most useful all-around is the LinRegTTest on the STAT TESTS menu. We will also study multiple regression. TI-83 and -84 calculators do not have a built-in function for this process, so on those calculators we will use a program called **A2MULREG,** included on the CD-ROM, in this work. TI-89 calculators do have built-in multiple regression capabilities.

All of the TI-calculators compute the correlation coefficient as a part of linear regression, so we will not discuss that separately.

SIMPLE LINEAR REGRESSION AND CORRELATION

EXAMPLE Old Faithful: Rangers at Yellowstone Park use (among other things) the duration of the last interval to predict the time between eruptions of the Old Faithful Geyser. We want to explore this relationship. The first step in exploring any relationship is to plot the data.

Duration (x)	240	120	178	234	235	269	255	220
Interval (y)	92	65	72	94	83	94	101	87

Scatter Plot

Enter the data into L1 (Duration) and L2 (Interval).

Set up Plot1 for a scatter plot as at right.

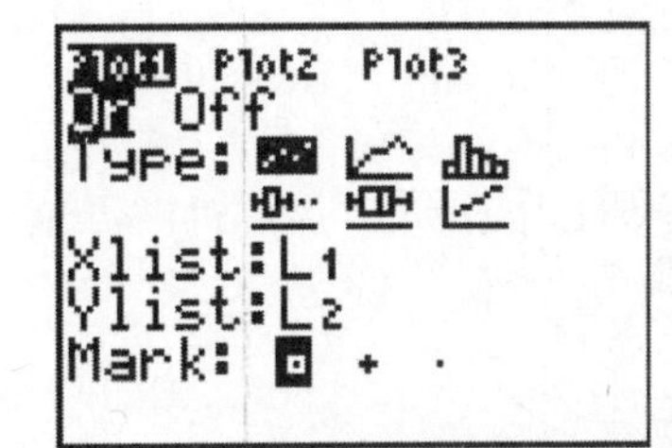

Press [ZOOM] [9] ([F5] on an 89 after the plot has been defined) for the plot.

The scatter plot seems to reveal a pattern indicating eruptions with longer durations are followed by a longer interval to the next eruption. The plot is also relatively linear.

Linear Correlation Coefficient, r

EXAMPLE Old Faithful: Using the data, find the value of the linear correlation coefficient r.

Press [STAT], arrow to TESTS and select LinRegTTest from the menu (the option number varies with calculator model, but is near the bottom of the list). Set up the screen as at right. On an 83 or 84, paste in Y1 from [VARS] [▸] [1] [1] to store the regression equation for further use. (On an 89, use the right and down arrows to select a regression equation function.)

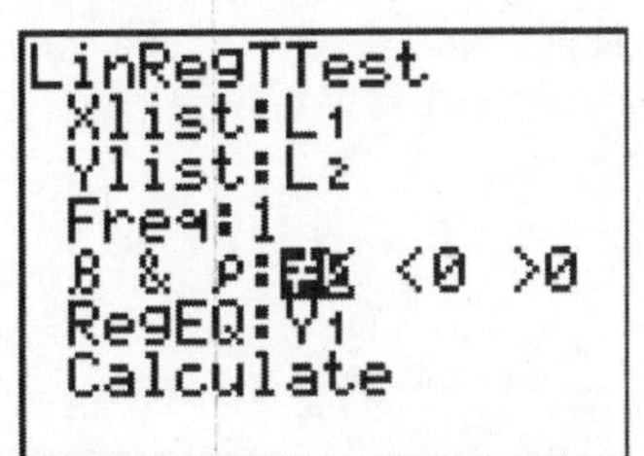

Highlight `Calculate` in the last line and press [ENTER] for the first of two screens of output. We are not interested in the t-statistic and its p-value at this point, so press the down arrow to scroll to the bottom.

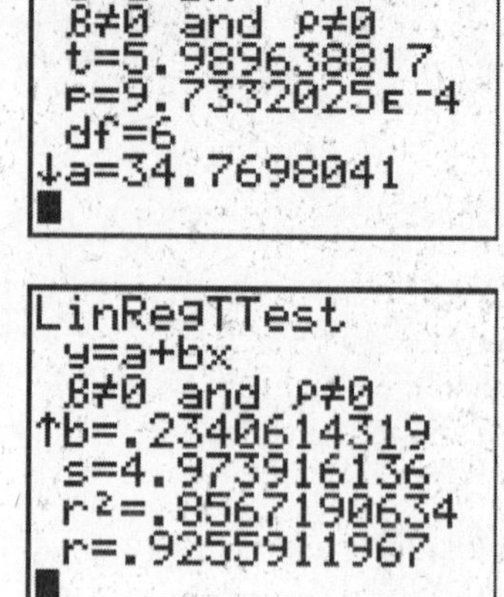

Note that the last line of output gives $r = .9255911967$. This indicates the linear relationship between these two variables is positive (as we saw in the graph) and strong (because r is close to 1).

Note: There are other ways of calculating r on TI calculators, but this method is more useful all around as it provides a lot of other results which will be of interest.

Explained Variation r^2 (Coefficient of Determination)

EXAMPLE Old Faithful: Referring to the Old Faithful data, what proportion of the variation in time to the next eruption can be explained by the duration of the last eruption?

In the screen above, we see that $r^2 = .8567...$, so about 85.7% of the variation in time to the next eruption can be explained by the variation in the duration of the last eruption.

Formal Hypothesis Test of the Significance of r

EXAMPLE Old Faithful: Using the data above, test the claim that there is significant linear correlation between duration of the last eruption and time to the next eruption . This test of H_0: $\rho = 0$ versus H_1: $\rho \neq 0$. is exactly the same as a test of H_0: $\beta_1 = 0$ versus H_1: $\beta_1 \neq 0$.
Using the first output screen at the top of this page, we see the test statistic $t = 5.9896$ and the p-value = 9.733E-4 = 0.000973 < .05, indicating significant positive linear correlation (and a non-zero slope).

Regression

EXAMPLE Old Faithful: For the data we have been working with, find the regression equation of the straight line that relates duration and time to the next eruption.

The equation has the general form $\hat{y} = b_0 + b_1x$. Your calculator uses the form $\hat{y} = a + bx$, so $a = b_0$ and $b = b_1$. You can look at the output above and see that a = 34.770 and b = 0.234 (rounded to three places as suggested in the text). Thus, your equation is $\hat{y} = 34.770 + 0.234x$. Be careful – many instructors will insist (like the author of this manual) that you report your regression equation *in terms of the actual variables with units* and *not* x and y. In this case, I would report "Time to next eruption (min) = 34.770 + 0.234*Duration of last eruption (sec)."

When we performed the calculation, we told the calculator to store the regression equation as `Y1`. You can check this by pressing [Y=]. ([◆][F1] = [Y=] on a TI-89.) It would be nice to see how the line passes through the data points.

We already defined the scatter plot of the data. If all other plots are turned off, you can press [ZOOM] [9] to see the regression line plotted through the scatter plot of points. (On a TI-89, go back to the plot set-up screen ([F2] from the Statistics/List Editor, and press [F5] to display the plot).

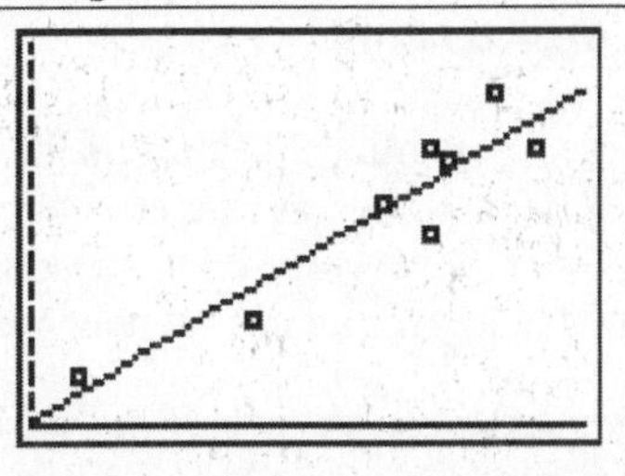

Predictions

EXAMPLE Old Faithful: Using the sample data, we found that there is a significant linear correlation between the duration of the last eruption of the geyser and the time interval to the next eruption. You also found the regression equation. Now suppose that a particular eruption lasted 180 seconds. Use the regression equation to predict the time interval after the eruption (time to the next eruption).

With the regression line and scatter plot displayed as above, press [TRACE] ([F3] on an 89) followed by [▼] to hop from a data point to the regression line. You can use the left and right arrows to try to locate $x = 180$ on the line, but it will be difficult.

Type in 180. A large X=18Ø will show in the bottom line. Press [ENTER]. We see the predicted y value is 76.9 minutes to the next eruption.

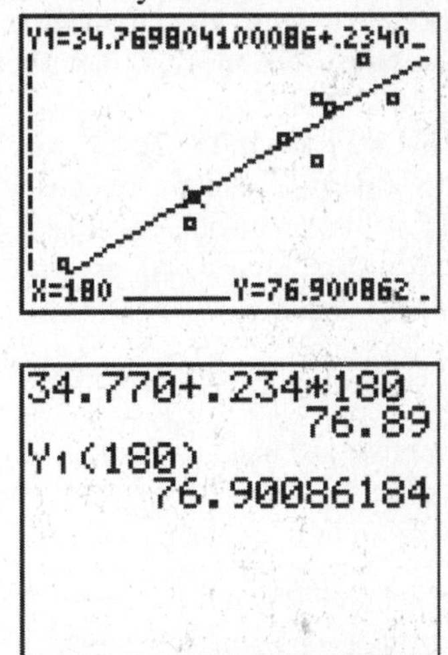

You can also calculate the predicted $\hat{y}$ on the Home screen by typing in the regression equation as in the top line of my screen or by pasting Y1 ([VARS] [▶] [1] [1]) to the Home screen and then typing (180) and pressing [ENTER], as in the last line of the screen. Notice a small difference due to rounding. (On a TI-89, locate the y1 variable from the [2nd][-] (VAR-LINK) screen.

Residuals Plots

Before proceeding with any more inference on a regression, it is good practice to examine a plot of the residuals (which represent unexplained variation around the line). We do this because these plots help assess whether the line is a good fit to the data. Having subtracted out the linear trend will magnify any remaining patterns. Ideally, when plotted as a scatter plot against the original x variable, one will see "pure, random scatter." Any noticeable pattern in the residuals plot means that the linear regression is *not* a good model for that set of data.

Define a scatter plot using the original x (predictor) variable, and using RESID as the y variable. Locate the list name RESID on the [2nd][STAT] (LIST) menu of list names. The TI calculator computes the residuals automatically when performing the regression. If you are using a TI-89, you will actually see these as a new list in the list editor after calculating the regression. Find the list name on the [VAR-LINK] screen in the STATVARS folder. Press [2] ('r' to move to the correct portion of the list).

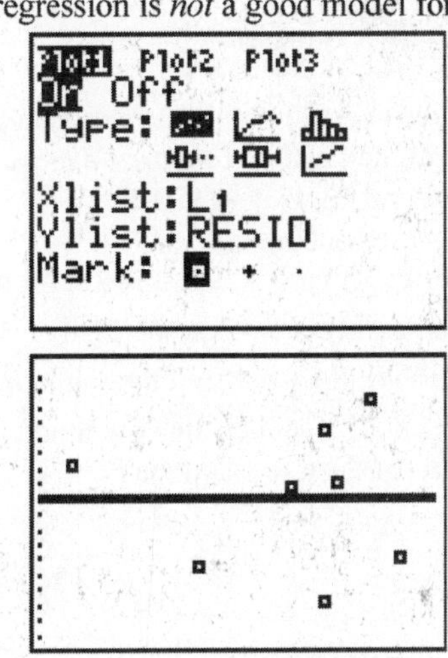

Press [ZOOM][9] ([F5] on an 89) to display the graph. Here we see pretty much random scatter, so the linear regression is adequate. One cautionary note: the calculator will try to display everything it can – any lines seen in this plot (except for y = 0 in the center) are not part of the plot!

Prediction Intervals

We met confidence intervals for the mean of a distribution earlier. So far we have used our regression equation to find a point estimate of the response for a given x – but there is still variation around the linear model (represented by the residuals). We'd like a prediction *interval* for a particular value of the x variable – one that gives an idea of how precise our estimate is.

EXAMPLE Old Faithful: For the Old Faithful data, we just found that the best prediction for $x = 180$ (a 180 second (3 minute) eruption) is $\hat{y} = 76.9$ minutes to the next eruption. Find a 95% prediction interval for the time to the next eruption if the last eruption lasted 180 seconds.

TI-83/84 Procedure

The only way to do this is to calculate the interval on the Home screen using the formula. From Table A-3 of the text, we find the value $t_{\alpha/2} = 2.447$ using 8 - 2 = 6 degrees of freedom. The other quantities needed in the formula will be stored in the calculator after you have performed the LinRegTTest. You can paste them in from the VARS menu as described below.

First we store 180 as X, 2.447 as T and Y1(X) as Y. Here Y1(X) is the value we get when we input X = 180 into the regression equation stored in Y1. This can be done as separate commands as shown, or as one command with the individual parts separated by a colon ([2nd][.]).

```
180→X
                180
2.447→T
              2.447
Y1(X)→Y
        76.90086184
```

Press [ENTER] to reveal the prediction value 76.9.

Now we calculate the margin of error E using the formula from the text. Values used in the formula are stored in the Statistics submenu of the VARS menu. You can find and paste in:

s at [VARS] [5] [▶] [▶] [▶] [0] (in the Test menu)
n at [VARS] [5] [1] (in the XY menu)
x̄ at [VARS] [5] [2] (in the XY menu)
Σx² at [VARS] [5] [▶] [2] (in the Σ menu)
and **Σx** at [VARS] [5] [▶] [1] (in the Σ menu)
Press [ENTER] to reveal that E = 13.434.

```
T*s*√(1+1/n+n(X-
x̄)²/(n*Σx²-(Σx)²
))→E
        13.43403617
```

Note: As you work, look carefully at the quantities in the statistics submenus, so you can familiarize yourself with what is available to you and where it is located.

Now we start with our prediction value Y from the regression equation itself and add and subtract the margin of error to form the 95% confidence interval $63.47 < y < 90.33$.

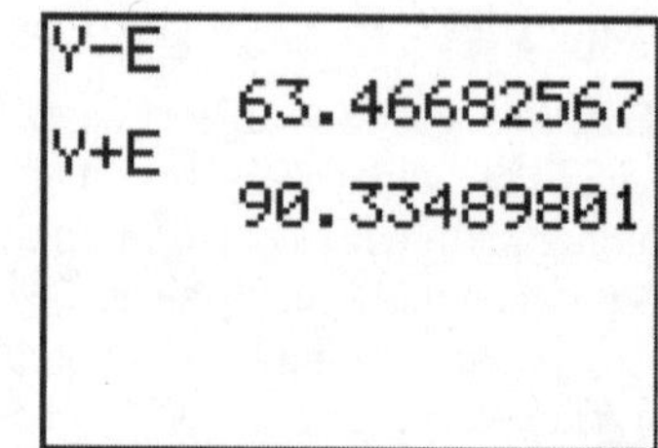

Note: The prediction interval for different x-values and different confidence levels can be easily calculated using the last-entry feature ([2nd] [ENTER]). Simply recall the top input lines beginning with where we stored our x value and change the inputs as needed and press [ENTER]. Then recall the equation to calculate E.

TI-89 Procedure

The TI-89 calculator has a built-in function to compute confidence intervals for both the slope and mean response at a particular x value, as well as prediction intervals for a particular response at a particular x value. From the `Ints` menu, select option `7:LinRegTInt`. Input the names of the X List and Y list you are using, and indicate that each entry is a single observation. Select a regression equation to store the results. With the right arrow, you can select to compute an interval for either the slope or the Response. If for a response, you will then be allowed to enter the particular x value. Use the down arrow to enter the particular confidence level for your problem.

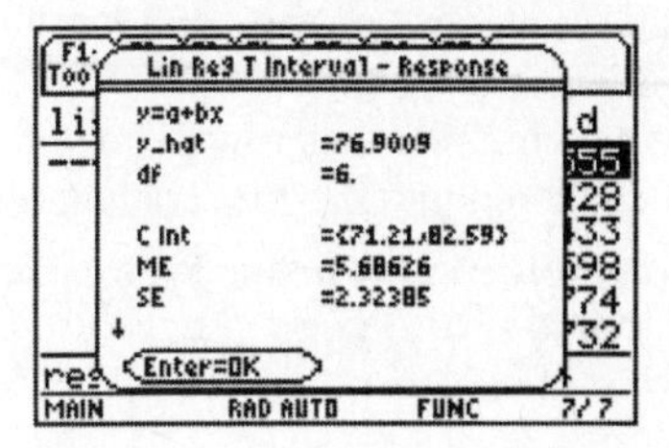

After pressing ENTER, you will see the first portion of the results. Here, we have y-hat, the value from the regression equation, and the confidence interval for the *mean* time to next eruption for *all* eruptions that last 180 seconds.

Using the down arrow, we find the prediction interval for a *particular* eruption that lasted 180 seconds. We are 95% confident that for a particular eruption that lasted 180 seconds, the time to the next eruption will be between 63.47 and 90.33 minutes. Completing the output screen are the coefficients of the regression, r^2, r, and the x value used.

MULTIPLE REGRESSION

The TI-83 and -84 do not have built-in multiple regression capabilities (regression using more than one predictor variable). TI-89 calculators do have built-in multiple regression. For the 83/84 calculators, we will use a program written for this purpose called **A2MULREG**. This program comes on the CD-ROM which accompanies your text. See the Appendix for details. Program **A2MULREG** requires that the data set be stored in matrix [D] on your calculator. We will walk through this process below. One important restriction is that the dependent variable (Y) must be stored in the first column of [D]. This is not the case when using most statistical software such as Minitab (or the TI-89).

EXAMPLE Old Faithful: We have already examined the relationship between duration of an eruption and the time to the next eruption. Perhaps if we add the height of the eruption (as a measure of how strong it was), perhaps we can improve our estimates of the time to the next eruption.

Time to Next	Duration	Height
92	240	120
65	120	110
72	178	125
94	234	120
83	235	140
94	269	120
101	255	125
87	220	150

TI-83/84 Procedure:

The data can be imported from another TI-83 Plus or from a computer. The following method is for entering the data from a keyboard. Press [2nd] [x^{-1}] [▶] [▶] to call the `MATRIX` menu and to choose the `EDIT` submenu. (If you are using a regular TI-83 calculator, it has a [MATRX] button to start the `MATRIX` menu.) Go down the screen to highlight `[D]` and press [ENTER]. You must first tell the calculator the size of the matrix you will be using. In the example below, we have 8 observations, and three variables, so the matrix is 8 x 3 (8 rows and 3 columns).

```
MATRIX[D] 8 ×3
[ 0     0     0  ]
[ 0     0     0  ]
[ 0     0     0  ]
[ 0     0     0  ]
[ 0     0     0  ]
[ 0     0     0  ]
[ 0     0     0  ]
1,1=0
```

Now enter the data row by row; however, the dependent (Y) variable *must* be entered into the first column of the matrix. Start with the cursor in the first row, first column. Type each data set value followed by [ENTER]. Press [2nd][MODE] to exit the editor.

```
MATRIX[D] 8 ×3
[ 65    120   110 ]
[ 72    178   125 ]
[ 94    234   120 ]
[ 83    235   140 ]
[ 94    269   120 ]
[ 101   255   125 ]
[ 87    220   150 ]
8,3=150
```

With the data stored in [D], we now use the program **A2MULREG** to verify the Minitab output given in the text for the three most important components. When the program is first started, you will see this screen. An indication that the calculator is paused is given by the scrolling dots in the upper right corner.

```
DATA IN MAT [D]
COL Y,X1,X2,..XN
Y MUST BE IN THE
1ST COL OF [D].

[A],[B],[C],[D],
[E]+[F] USED.
```

Press [ENTER] to get the menu screen. Press [ENTER] to select a multiple regression.

```
MULT REG+CORR
1:MULT REGRESSIO
2:CORR MATRIX
3:QUIT
```

You will then be prompted for the number of independent (predictor) variables. You are now asked which columns the independent variables are in.

```
HOW MANY IND VAR
?2
COL. OF VAR.
               1
?2
COL. OF VAR.
               2
?3
```

Here is the first screen of output. What we see first is the analysis of variance output for the general usefulness of the model. This F test tests the hypotheses H_0: all $\beta_I = 0$ versus H_1: not all $\beta_I = 0$. In this case, the F statistic is F = 16.91 with p-value p = 0.006, so we conclude that not all the coefficients of our predictor variables are 0 (at least one is significantly related to the time to next eruption). We also see r^2 for the entire model, the r^2 adjusted for multiple predictor variables, and the standard deviation of the points around the regression "surface" (it's not a line in multiple dimensions).

```
   DF SS
RG 2  897.694939
ER 5  138.305060
     F=16.23
     P=.007
 R-SQ=.8665
 (ADJ).8131
S=5.259373737
```

Press [ENTER] again to find the estimates for the coefficients, along with their individual t statistics and p-values.

The multiple regression equation is given as

Time to Next Eruption = 45.105 + .245*Duration – 0.098*Height.

We are also given the t-statistics for testing the individual coefficients and the p-values for the test of H_0: $\beta_I = 0$ versus H_1: $\beta_I \neq 0$. Notice that for the third column (the variable height) the t-statistic is small and the p-value large, so this coefficient is not significantly different from 0.

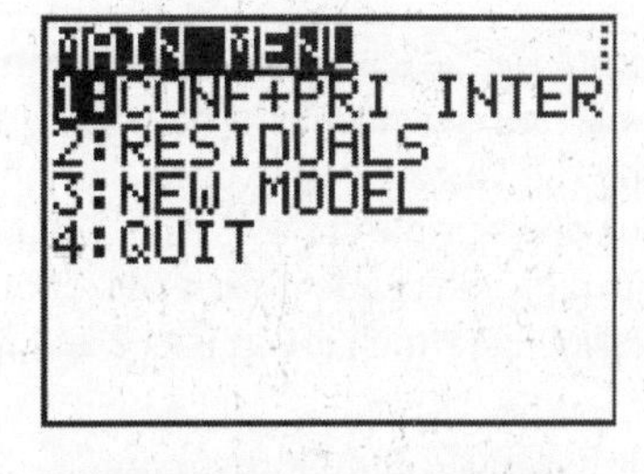

Press [ENTER] again and you will be presented other choices to continue the analysis. The first option will compute confidence and prediction intervals for combinations of the predictor variables (the program will tell you the degrees of freedom for the model – you will need to supply the appropriate t-multiplier); the second will do residuals plots. The submenus and prompts for these options are self-explanatory.

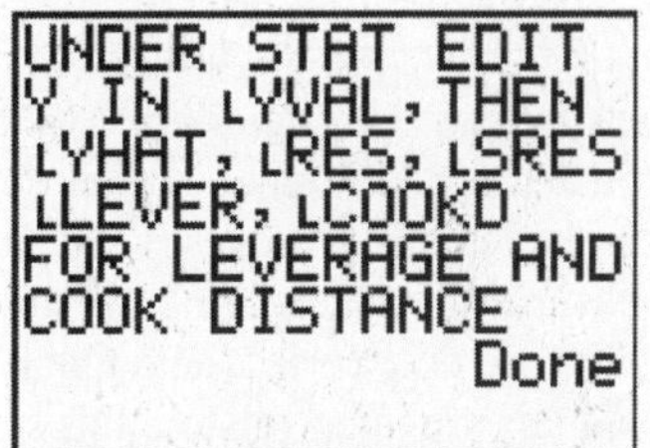

When you select the QUIT option, you will see this message screen. The normal lists L1 through L6 have been replaced with six new lists – the actual y-values in the problem, the y-hats (estimated from the model), residuals and standardized residuals (residual divided by s), and two other lists Leverage and Cook's Distance. The last two can be used for diagnostic purposes. Both Leverage and Cook's distance measure the impact of each data point in computing the model. Larger values indicate the point was more influential.

YVAL	YHAT	RES
92	90.063	1.9374
65	63.654	1.3462
72	76.369	-4.369
94	90.56	3.4402
83	88.839	-5.839
94	99.122	-5.122
101	95.206	5.7941

YVAL(1) =92

This is what the Statistics Editor will look like after running the program. To recover the normal statistics lists, use option 5:SetUpEditor from the main [STAT] menu.

Correlation Matrix

EXAMPLE: What is the best regression equation if only a single independent variable is to be used?

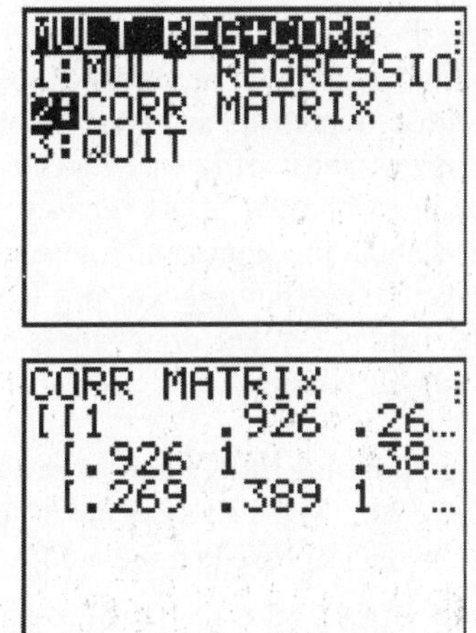

When you ran Program **A2MULREG**, the first menu screen gave you the option of calculating the correlation matrix. Begin running the program again. This time highlight the option of calculating the correlation matrix.

Note: Pressing [ENTER] from the Home screen will restart the program if the last thing done was to quit it.

Press [2] and wait patiently as this calculation can take some time, depending on the number of variables and data points. The partial output is given at right. Looking at just the first column, we can see that the last variable, Height, has a small linear correlation with the first variable, Time to Next Eruption. This explains why its coefficient was not significantly different from 0.

TI-89 Procedure

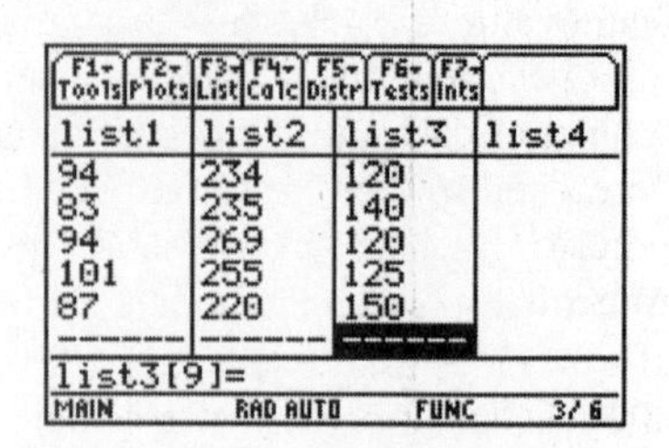

Enter the data into the normal statistics lists, as you would for anything else. Here, I have entered the Time to Next Eruption in `list1`, the Duration of the Last Eruption in `list2`, and the Height of the last Eruption in `list3`.

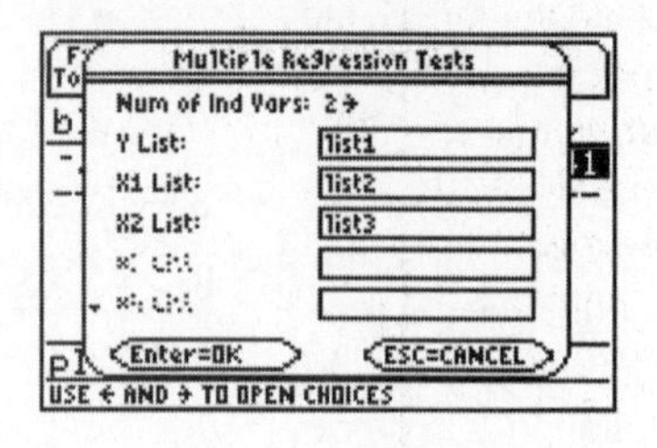

From the `Tests` menu, select option `B:MultRegTests`. You first must tell the calculator how many independent variables you will use. Use the right arrow key to open the submenu, and use the down arrow to find the appropriate choice. This will change the number of boxes available to specify independent variables. For our problem, there are 2 independent variables, which are in lists 2 and 3.

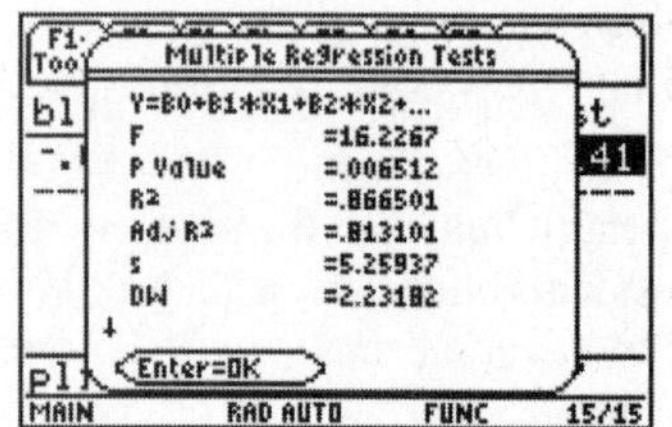

Press ENTER for the first portion of output. This gives the F statistic and p-value for the overall "utility" of the model. This tests the hypotheses H_0: $\beta_I = 0$ versus H_1: not all $\beta_I = 0$. It also gives the values of r^2 and r^2 adjusted for additional variables in the model. The standard deviation of the residuals is given, as is the Durbin-Watson statistic , DW, which measures the amount of correlation in the residuals and is useful for data which are time series (data that has been collected through time). If the residuals are uncorrelated, this statistics will be about 2 (as it is here); if there is strong positive correlation in the residuals, DW will be close to 0;
if the correlation is strongly negative, DW will be close to 4. Since these data are not a time series, DW is meaningless for our example.

Pressing the down arrow several times we find the components for regression and error which are used in computing the F statistic. The F statistic for regression is the MS(Reg)/MS(Error) where the Regression Mean square functions just like the treatment (factor) mean square in Analysis Of Variance (ANOVA).

Finally we see some of the entries in new lists that have been created. The complete lists will be seen when ENTER is pressed. `Blist` contains the estimated intercept and coefficients; `SE list` is the list of standard errors for the coefficients which can be used to create confidence intervals for true slopes; `t list` gives values of the *t*-statistics for hypothesis tests about the slopes and intercept; `P list` gives the p-values for the tests of the hypotheses H_0: $\beta_i = 0$ against H_A: $\beta_i \neq 0$. If the assumed alternate is 1-tailed, divide these p-values by 2 to get the appropriate p-value for your test.

After pressing ENTER we see several new lists that have been added into the editor. `Yhatlist` is the list of predicted values for each observation in the dataset based on the model ($yhat_i = b_0 + b_1x_{1i} + b_2x_{2i}$ in this model); `resid` is the list of residuals $e_i = y_i - yhat_i$. `Sresid` is a list of standardized residuals obtained by dividing each one by S, since they have mean 0. If the normal model assumption for the residuals is valid, these will be N(0, 1).

Pressing the right arrow we find yet more lists. Leverage is a measure of how influential the data point is. These values range from 0 to 1. The closer to one, the more influential (more of an outlier in its x values) the point is in determining the slope and intercept of the fitted equation. Values greater than 2p/n, where n is the number of data points and p is the number of parameters in the model, are considered highly influential. Here, $n = 8$ and $p = 3$, so any value grater than 0.75 will designate an observation as highly influential. This indicates in our example that the second data point (the one with a 65 minute time to next eruption) is influential.

Cook's Distance in the next column is another measure of the influence of a data point in terms of both its x and y values. Its value depends on both the size of the residual and the leverage. The ith case can be influential if it has a large residual and only moderate leverage, or has a large leverage value and a moderate residual, or both large residual and leverage. To assess the relative magnitude of these values, one can compare them against critical values of an F distribution with p and $n - p$ degrees of freedom or use menu selection `A: F Cdf` from the [F5] (`Distr`) menu. The largest value in the list is .5311, so that point is not highly influential.

F1 Tools | F2 Plots | F3 List | F4 Calc | F5 Distr | F6 Tests | F7 Ints

cookd	blist	selist	tlist
.01746	45.105	19.411	2.3236
.39574	.24464	.04486	5.4531
.08832	-.0983	.16232	-.6053
.05995	------	------	------
.15641			
.53111			

tlist={2.323620491512,5.4...

MAIN RAD AUTO FUNC 14/15

After pressing the right arrow still more, we find the last of the output lists.
`Blist` is the list of coefficients. We finally see the fitted regression equation:
$\textit{Time to Next Eruption} = 45.105 + 0.245 * \textit{Duration} - 0.098 * \textit{Height}$. We interpret the coefficients in the following manner: Time to next eruption increases .245 minutes for each second of duration for the last eruption, for eruptions of the same height, while time to next eruption decreases 0.098 minutes for each foot of height in the last eruption, for eruptions of the same duration.

The next column contains the standard errors of each coefficient. These can be used to create confidence intervals for the true values using critical values for the t distribution for $n - p$ degrees of freedom. Finally we see the t statistics and p-values for testing H_0: $\beta_i = 0$ against H_A: $\beta_i \neq 0$. These suggest the coefficient of height is not significantly different from 0; in other words, duration is a much more determining quantity for the time to the next eruption of Old Faithful.

Confidence and Prediction Intervals

The TI-89 can also calculate automatically confidence intervals for the mean response for a combination of the predictor variables, as well as prediction intervals for individual new observations. Use option 8:MultRegInt from the [F7] Ints menu. Specify the values for the variables in the X Values List box enclosed in curly braces ([2nd][(] and [2nd][)]) and separated by commas.

How do I get rid of those extra lists?

From the Statistics list editor, press [F1] (`Tools`) and select option `3:Setup Editor`. Performing this action (leave the input box blank) recovers any deleted lists and deletes any calculator-generated lists.

11 Multinomial Experiments and Contingency Tables

In this chapter we will use the TI-83 Plus STAT Editor as a spreadsheet to calculate the χ^2 test statistic used in the goodness of fit test for multinomial experiments. If you are using a TI-84 or -89, this is a built-in function. We will also use a built-in test from the STAT TESTS menu to do contingency table analysis.

MULTINOMIAL EXPERIMENTS: GOODNESS OF FIT TESTS

EXAMPLE Last Digits of Weights: The table below is a recreation of part of Table 11-2 in your text. It contains a frequency analysis of the last digits of weights obtained from 80 randomly selected students. If the students had actually been weighed, instead of being asked their weight, we would expect each digit to appear equally. Test the claim that reported weights do not have this property. Use significance level α = 0.05.

Last digit	0	1	2	3	4	5	6	7	8	9
Expected Frequency	.10	.10	.10	.10	.10	.10	.10	.10	.10	.10
Observed Frequency	35	0	2	1	4	24	1	4	7	2

Calculating Expected Frequencies (all models)

Put the observed frequencies into L1 and the expected proportions into L2.

Highlight L3 and type L2*sum(L1 as in the bottom of the screen (Find the sum(function at [2nd] [STAT] [►] [►] [5], option 5 on the LIST MATH menu.)

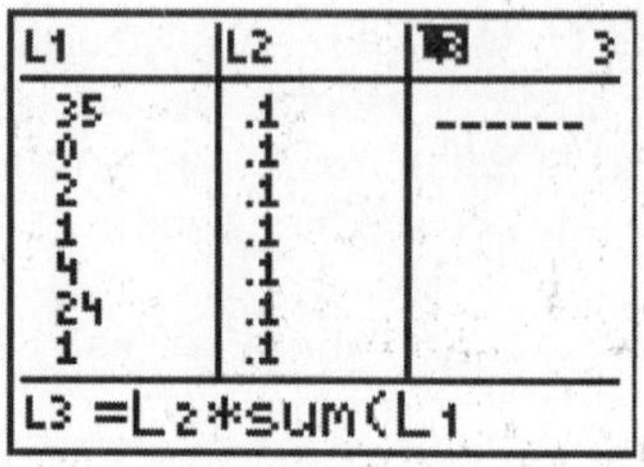

Press [ENTER] for the expected frequencies (E) in L3. (If these values are given or are easy to calculate as in this example, you can simply enter them into L3.)

Conducting the χ^2 Test – TI-83 Procedure

Highlight L4 and type (L1-L3)²/L3. Press [ENTER] for the contribution made by each of the digits to the overall chi-square statistic.

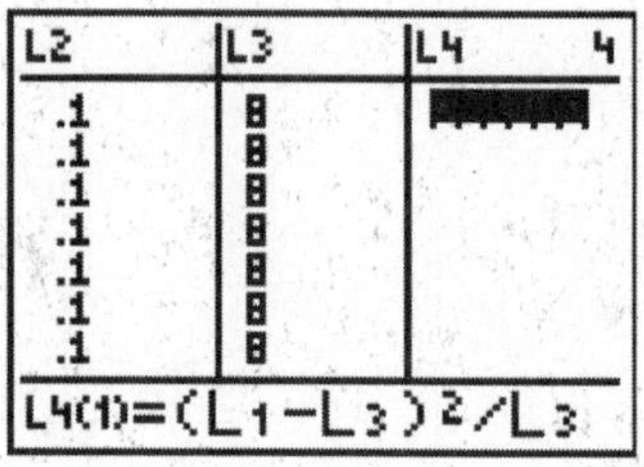

We can see that the largest contribution is made by an ending digit of 0. The next largest contribution is from 5. This seems to confirm our nagging suspicion that most people report their weights in values ending in 0 or 5.

Press [2nd] [MODE] to Quit and return to the Home Screen. Sum the elements of L4, using the Sum(command from the LIST MATH menu (as above). We see the chi-square statistic is 156.5.

Computing the p-value of the χ^2 Statistic

Press [2nd] [VARS] [7] to paste χ^2cdf(on the home screen. Then type 156.5,E99,9 and press [ENTER] for a p-value as shown. (You have asked for the probability of a test statistic greater than 156.5 in a chi-square distribution with df = 9.) Since the p-value is essentially 0 and much smaller than the significance level $\alpha = 0.05$, we would reject the null hypothesis that individuals report their weight accurately.

Conducting the χ^2 Test – TI-84/89 Procedure

Both of these models have a built-in function to compute the test statistic and find the p-value of the test. In both cases, select χ2GOF from the Stat Tests menu. Both models ask for the observed and expected lists, as well as the degrees of freedom (categories – 1) and give the option to either simply calculate or draw the results.

Pressing [ENTER] will display the results. The TI-89 will also create a new list with the components of the χ2 statistic for examination purposes.

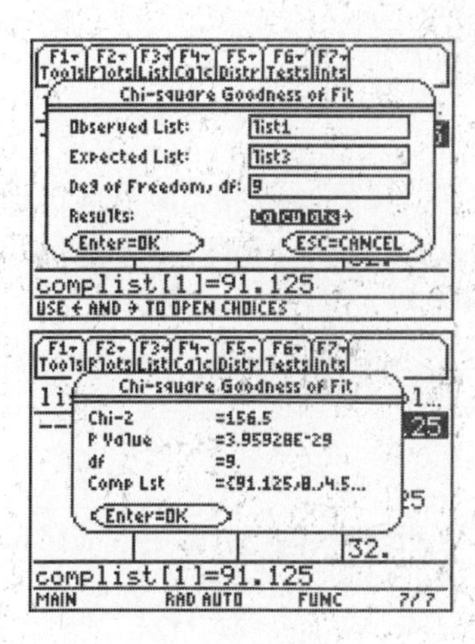

Observing Differences Graphically

To graphically compare the observed frequencies with the expected frequencies, put the integers 0 to 9 in L5.

Set up Plot1 and Plot2 to be connected scatterplots using the digits in L5 as the Xlist and the observed and expected frequencies as the Ylist. As illustrated, use a different mark for the observed frequencies in L1 and the expected frequencies in L3. Press [ZOOM] [9] [TRACE] to view the plot.

Note: The observed proportions and expected proportions could have been plotted as in Figure 10.6(a) of the text, but the graph would look the same-only the y-axis scale would change.

This is the finished plot. Notice last digits 0 and 5 are over-reported, all others (except 8) are underreported.

EXAMPLE Detecting Fraud: Refer to the table below which is a recreation of part of Table 11-1 in your text. It contains a frequency analysis of the leading digits on 784 checks. Test the claim that the digits do not occur with the frequency stipulated by Benford's Law. Use significance level $\alpha = 0.01$.

Lead digit	1	2	3	4	5	6	7	8	9
Expected % Frequency	30.1	17.6	12.6	9.7	7.9	6.7	5.8	5.1	4.6
Observed Frequency	0	15	0	76	479	183	8	23	0

Here, I have entered the observed frequencies into L1, the expected frequency (percent, as a decimal) in L2, and computed the expected frequency (count) in L3, just as was done above (multiply L2 by the sum of L1).

```
L1     L2     L3
0      .301   235.98
15     .176   137.98
0      .126   98.784
76     .097   76.048
479    .079   61.936
183    .067   52.528
8      .058   45.472
L3(1)=235.984
```

On my TI-84 calculator, I have completed the input screen, similarly to that done above on a TI-89. This time, I have chosen the Draw output option.

```
χ²GOF-Test
 Observed:L1
 Expected:L3
 df:8
 Calculate Draw
```

The distribution curve displayed is not symmetric. Further, (to 4 decimal places) the p-value of this test is 0 (so no area is shaded). The computed test statistic is 3651.0354. We have shown that this sample of checks does not follow Benford's Law. Fraud may be involved!

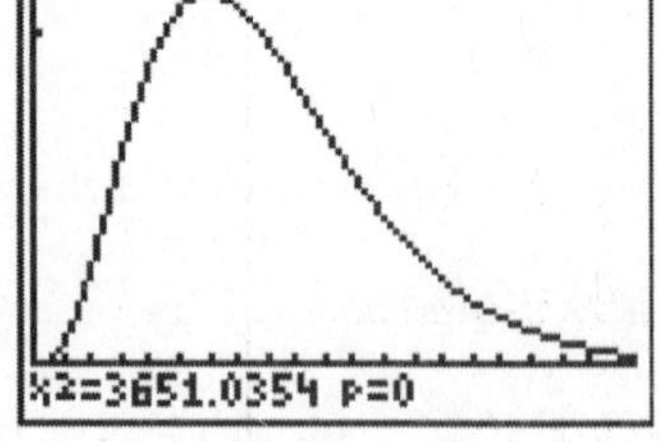

CONTINGENCY TABLES

Contingency tables are used to analyze count data for two types of relationships: homogeneity and independence. The test is the same for both – a $\chi 2$ test, but the setting and conclusions are very different. It is important to make the connection what type of test you are performing. A test of independence usually involves the same individuals or items categorized by two variables. We want to know whether there is an association (or not) between the two variables. A test of homogeneity involves a single variable observed at different times, or in different populations. Here, we want to know if the distribution of the variable is the same in each.

EXAMPLE Motorcycle Helmet Color: The data presented in table 11-5 represent observations from a retrospective study which sought to answer the question of whether helmet color is related to risk of injury for motorcycle riders. The data were for a control group (those randomly stopped by police), and a case group which were riders seriously injured or killed. In this test, we want to know if being injured in an accident is related to the color helmet.

	Color of Helmet		
	Black	White	Yellow/Orange
Controls	491	377	31
Cases	213	112	8

TI-83/84 Procedure

We first must enter the contingency table into a matrix. Press [2nd][x^{-1}] ([MATRX] on a regular TI-83) and arrow to EDIT. Select a matrix (usually [A]) and press [ENTER] to get the input screen. You first must tell the calculator the size of the matrix. Ours has two rows and 3 columns, so press [2][ENTER][3][ENTER] to set the dimensions. Now input the data (across rows) pressing [ENTER] after each. Press [2nd][MODE] to Quit the editor.

Press [STAT], arrow to TESTS, and select C:χ2-Test. The input screen is pretty obvious. Tell the calculator the name of the matrix with your observed counts, and where to store the matrix it will create with the expected counts. Choose either Calculate or Draw, as usual. If you want to change a matrix name, press [2nd][x^{-1}] to access the menu of matrix names.

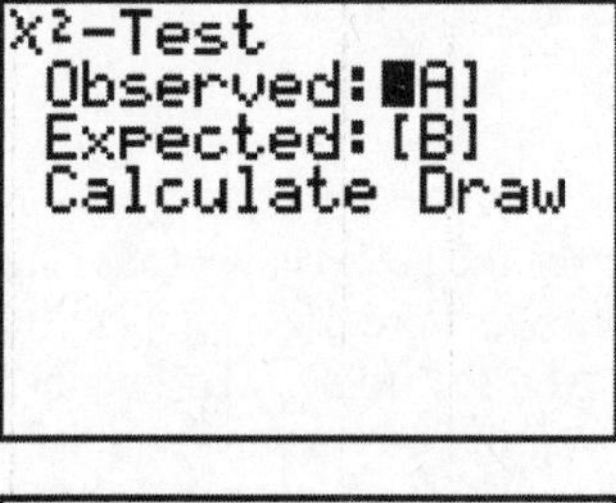

Here, our computed χ2 test statistic is 8.775, with p-value 0.0124. It appears that helmet color and whether a motorcycle rider has been seriously injured or killed are related.

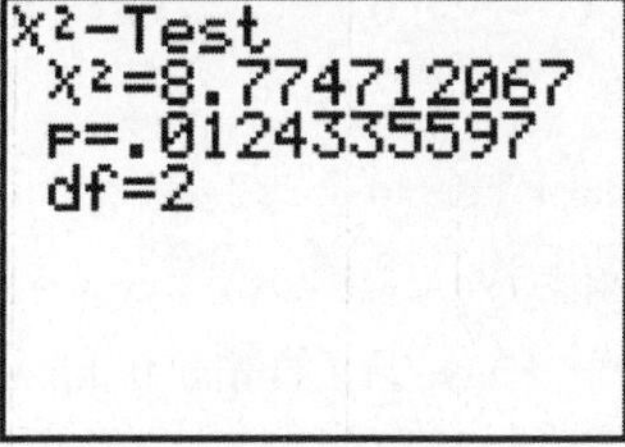

TI-89 Procedure

On a TI-89, we must first select the Data/Matrix Editor application. The first thing you will see is this. It asks whether you wish to open a current problem, an existing problem, or start a new one. I have selected to begin a new problem.

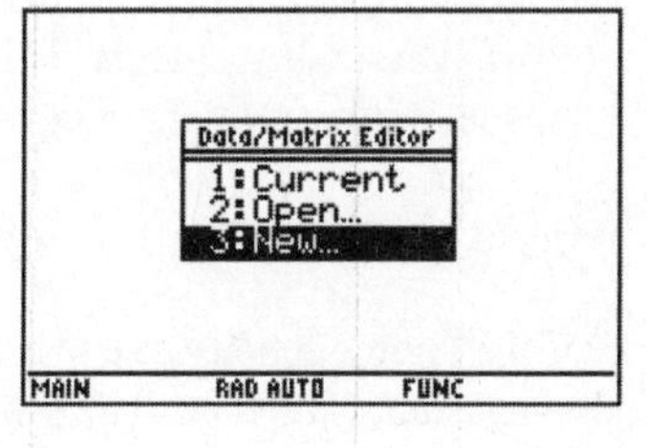

This next screen is used to define the type of entity you will be using. We want a matrix, so use the right arrow to expand the first type option box and use the down arrow to select (highlight) `Matrix`. You are asked which folder to store the matrix in, I recommend using `main`. Next, enter a name for the matrix, then the row and column dimensions. Press [ENTER] to proceed to the input screen.

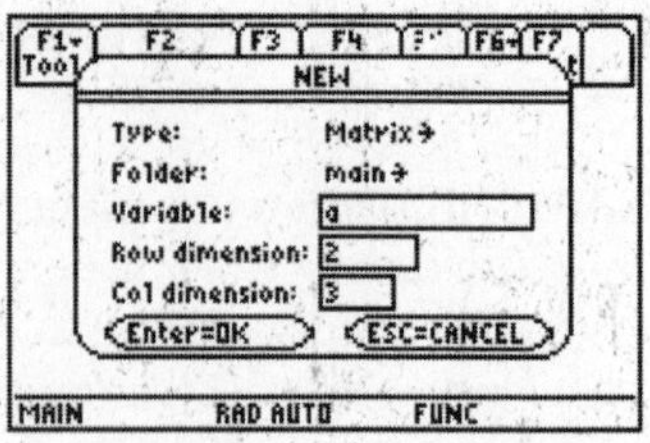

Simply type in the entries (across the rows) and press [ENTER] after each. To exit the editor, press [2nd][ESC].

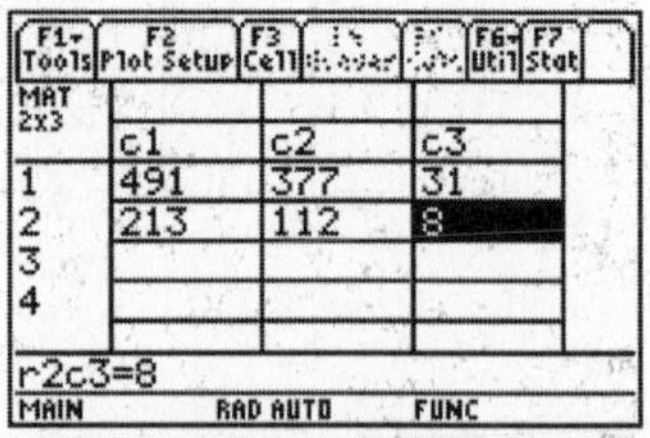

Return to the Stats/List Editor, and select option `8:Chi2 2-way` from the [F6] `Tests` menu. You are asked the name of the input matrix that you just created (access the name from [VAR-LINK] just as with list names), and have the option of changing the default names for the matrices of expected counts and components of the $\chi 2$ test statistic. Last, as usual, you have the option of `Calculate` or `Draw` for the results.

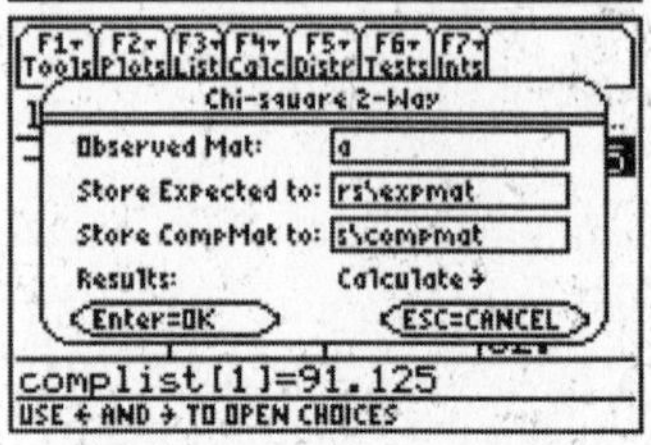

The results screen is at right. We see the value of the test statistic and its p-value along with degrees of freedom. We also see the first few entries in both the Expected Value matrix and Components matrix.

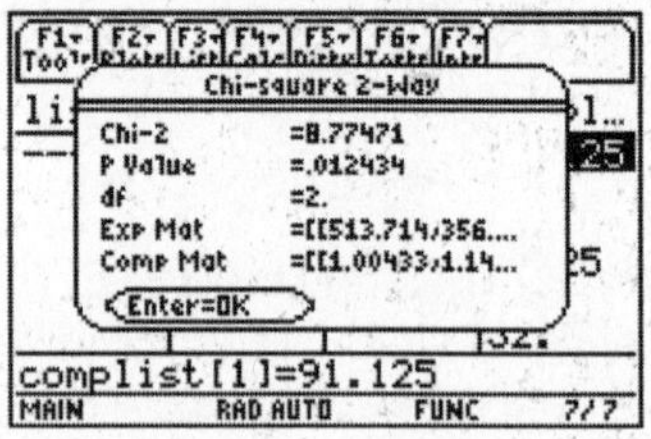

To see the actual expected values or components matrices go to the home screen and press [2nd][-] ([VAR-LINK]). Both of these are stored in the `Statvars` folder. To find them more easily, highlight `Main` and press ◀ to collapse the entries in that folder. Move the cursor to highlight `Statvars`. If needed, press ▶ to expand that list. Locate the entry `compmat` and press [ENTER] to transfer the name to the home screen entry area, then press [ENTER] again to display the matrix. The largest contributors to the $\chi 2$ test statistic are for those injured with black or white helmets.

Here, I have in a similar fashion displayed the entries in `expmat`, the matrix of expected counts. Comparing these to the matrix of original data, it seems that riders with black helmets are involved in serious accidents (213) more often than expected (190.3), and those in white helmets are involved in serious accidents (112) less often than expected (132.2). Perhaps visibility of the helmet is a contributing factor.

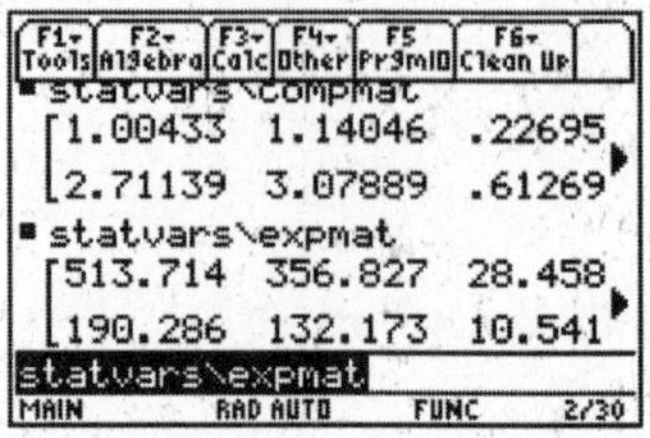

Tests of Homogeneity

In a test of homogeneity, we test the claim that different populations have the same proportions of some characteristic of interest. The procedure is the same as for tests of independence; one needs to be careful to identify the difference between the two.

EXAMPLE Influence of Gender: Does the gender of an interviewer affect the response given? A sample of 1200 men was asked if they agreed with the statement "Abortion is a private matter that should be left to the woman to decide without government intervention." The data in table 11-6 are reproduced below.

	Gender of Interviewer	
	Male	Female
Agree	560	308
Disagree	240	92

I have entered the observed data into matrix [A], as shown at right.

```
MATRIX[A] 2 ×2
[ 560    308    ]
[ 240    92     ]

2,2=92
```

As detailed above, I have performed the $\chi 2$ test. The test statistic value is 6.529, with p-value 0.0106. We (at $\alpha = 0.05$) will reject the null hypothesis that the proportion of those agreeing is the same for both genders of interviewers.

```
χ²-Test
 χ²=6.529343179
 p=.0106109132
 df=1
```

If we examine the matrix of expected values and compare it to the matrix of observed values, we see that our male subjects agreed with the statement less often than expected when the interviewer was a man.

```
MATRIX[B] 2 ×2
[ 578.67  289.33 ]
[ 221.33  110.67 ]
```

WHAT CAN GO WRONG?

Expected cell counts less than 5.
Check the computed matrix of expected cell counts. If they are not all greater than 5 the analysis may be invalid.

Missing or misplaced parentheses.
When computing elements for the goodness-of-fit test on a TI-83, the parentheses are crucial.

Overusing the test.
These tests are so easy to do and data from surveys and such are commonly analyzed this way. The problem that arises here is that in this situation the temptation is to check many questions to see if relationships exist; but performing many tests on *dependent* data (the answers came from the same individuals) such as this is dangerous. In addition, remember that, just by random sampling, when dealing at $\alpha = 5\%$ we'll expect to see something "significant" 5% of the time when it really isn't. This danger is magnified when using repeated tests - it's called the problem of multiple comparisons.

12 Analysis of Variance

In this chapter we will use the TI calculators' built-in function for doing one-way ANOVA problems. We will use a program called **A1ANOVA** (included on the CD-ROM) to extend our capabilities to do two-way ANOVA problems for two factor designs with equal numbers of observations in each cell on TI-83 and -84 calculators. The TI-89 has a built-in function for this. The output for this program matches the Minitab ANOVA tables in the text. See the text for the proper interpretation for these tables.

ONE-WAY ANOVA

EXAMPLE Weights of Poplar Trees: Given the data on poplar tree weight under different conditions summarized in Table 12-1 in the text and a significance level of $\alpha = 0.05$, test the claim that the samples came from populations with means that are not all the same. The actual data values are found in Data Set 7 of Appendix B and in the table below.

None L1	.15	.02	.16	.37	.22
Fertilizer L2	1.34	.14	.02	.08	.08
Irrigation L3	.23	.04	.34	.16	.05
Fertilizer and Irrigation L4	2.03	.27	.92	1.07	2.38

Put the data in lists L1 to L4 as indicated in the table. (You can also import from the Data App.)

Plot the data. In this case, with four distributions, the TI-83 and -84 calculators cannot graph all four distributions at once. The TI-89 can. I have defined four modified boxplots to display the distributions of the tree weights for the various treatments. See Chapter 3 if you need to refresh your memory on these. Looking at the plots, No treatment and Irrigation seem to have similar distributions. Clearly, the combination of Fertilizer and Irrigation has the largest median. The value 1.34 in the Fertilizer category seems like it may be an outlier – with so few data points, it is hard to tell.

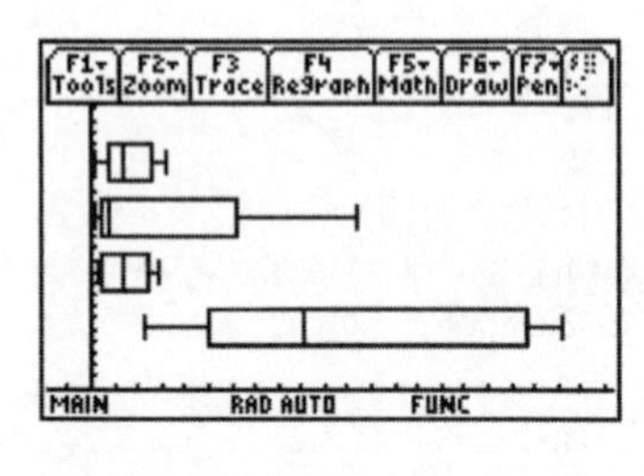

On a TI-83 or -84, press [STAT] ▶ ▶ ▲ [ENTER] to choose the last option on the STAT TESTS menu and paste ANOVA(on the Home screen. Type L1,L2,L3.L4 as at right. On a TI-89, select option C:Anova from the [F6] Tests menu. The input and outputs are so obviously alike what is shown, that no separate procedure to describe it is needed.

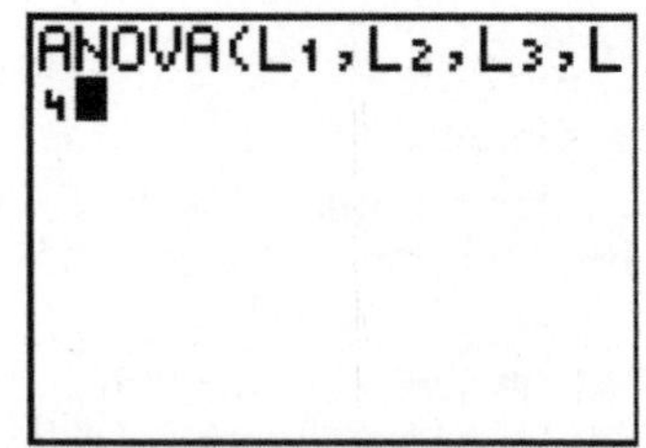

Press [ENTER] for the first portion of the results. We see the test statistic F = 5.731 and the p-value = 0.00735 just as seen in the text. The sum of squares for the factor (group) is shown along with its degrees of freedom and mean square. This small p-value confirms what our eyes already saw – the different treatments result in different weight trees. This does not, however, indicate which mean(s) is (are) different from the others – simply that one (or more) are different. Arrowing down, we see the same information given for the error sum of squares, and the (pooled) standard deviation.

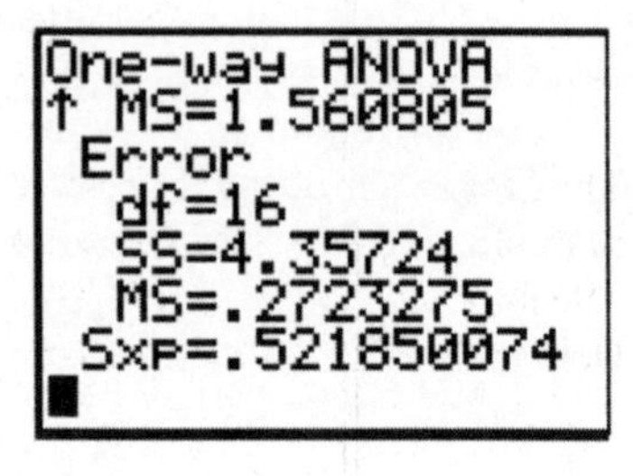

Note: The program **A1ANOVA**, introduced in the next section, gives the means and standard deviations of the raw data stored in the lists. The program also accepts sample summary statistics (means, standard deviations, and sample sizes) as an input option in addition to the raw data option. The TI-89 also accepts summary statistics and has the capability to perform two-way ANOVA.

Which mean(s) is (are) different?

Having rejected the null hypothesis, we would like to know which mean (or means) is (are) different from the rest. For a rough idea of the difference one can do confidence intervals for each mean (using the pooled standard deviation found in the ANOVA) and look for intervals which do not overlap, but this method is flawed. Similarly, testing each pair of means (doing six tests here) has the same problem: the problem of multiple comparisons. If we constructed four individual confidence intervals, the probability they all contain the true value is $0.95^4 = 0.8145$, using the fact that the samples are independent of each other.

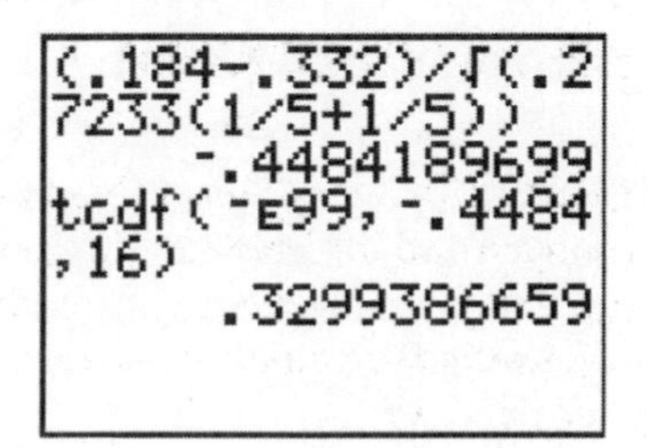

The Bonferroni method is one way to make the distinction. In this, we must be careful to control for the number of possible tests. `1-Var Stats` tells us that the mean for `L1` (no treatment) is 0.184 and the mean for `L2` (fertilizer) is 0.332. Are these means different, or similar? We will compute a t-test for equality of these means, and use a method to adjust the p-value to account for the possibility of several tests. The t-statistic is $t = \dfrac{\bar{x}_1 - \bar{x}_2}{\sqrt{MSE\left(\dfrac{1}{n_1} + \dfrac{1}{n_2}\right)}}$. I have computed it at right, and found the one-sided p-value for a test of equality of means. Multiply by two for a two-sided test, and we have p = .6599. Those two groups have similar means. (Continuing, we find that the group of trees that received both fertilizer and irrigation is the one that is different from the others – merely confirming what our eyes already told us.)

TWO-WAY ANOVA

The following example uses program **A1ANOVA** to perform two-way ANOVA. The program **A1ANOVA** is included on your CD-ROM, and from the book's website. To download the program requires TI-Connect software and a cable.

Two-Factor Design with an Equal Number of Observations per Cell

EXAMPLE Tree Growth: We extend the example of the previous section to consider another question: does location also make a difference in tree growth?

	No Treatment	**Fertilizer**	**Irrigation**	**Fertilizer and Irrigation**
Site 1	0.15 0.02 0.16 0.37 0.22	1.34 0.14 0.02 0.08 0.08	0.23 0.04 0.34 0.16 0.05	2.03 0.27 0.92 1.07 2.38
Site 2	0.60 1.11 0.07 0.07 0.44	1.16 0.93 0.30 0.59 0.17	0.65 0.08 0.62 0.01 0.03	0.22 2.13 2.33 1.74 0.12

The data in the table above will be stored in matrix [D] with 40 rows and 3 columns. The data values are all in the first column. The second column identifies the site associated with each data value (1=Site 1, 2=Site 2). The third column of [D] identifies the treatment associated (1 = None, 2 = Fertilizer, 3 = Irrigation, 4 = Both). Here are the step by step instructions.

Entering Data into Matrix [D]

Press [2nd] [x⁻¹] [▸] [▸] [4] to call the MATRX menu, the EDIT submenu and choose matrix [D] to edit. Type 40 [ENTER] 3 [ENTER] for 40 rows and 3 columns. Your matrix may not contain all zeroes, but that is fine as we will be typing over the values. Enter data row by row. Begin with 0.15 [ENTER] 1 [ENTER] 1 [ENTER]. Then continue typing in all the data values with their correct site and treatment identifier. Press [2nd][ESC] to exit the matrix editor.

```
MATRIX[D] 40×3
[ 1.07   1    4  ]
[ 2.38   1    4  ]
[ .22    2    4  ]
[ 2.13   2    4  ]
[ 2.33   2    4  ]
[ 1.74   2    4  ]
[ .12    2    4  ]
40,3=4
```

Running Program A1ANOVA

Press [PRGM] and highlight **A1ANOVA**. Press [ENTER] and prgm**A1ANOVA** is pasted to the Home screen. Press [ENTER] for the main menu screen. Press [3] to choose option 3, the 2WAY FACTORIAL.

```
ANOVA USES[D][E]
1:ONE-WAY ANOVA
2:RAN BLOCK DESI
3:2WAY FACTORIAL
4:QUIT
```

You will then see this instructional screen. The purpose of this screen is primarily to remind you of the input data requirements in case you should forget. Press [ENTER] another menu screen. This gives you two options: CONTINUE or QUIT. You could quit if the data were not already in matrix [D] as required.

```
EQUAL REPLICATES
DATA IN N*3 MAT
[D],1ST COL-DATA
2ND COL-A LEVELS
3RD COL-B LEVELS
LEVELS-INTEGERS
STARTING WITH 1.
```

Press [1] to continue since our data is stored in [D]. The first part of the ANOVA table is shown on the screen. For each factor (A is the factor used in column 2 of the input matrix, so site for our example; B is for the individual tree treatments), as well as interaction and error, the degrees of freedom and sum of squares is given. Mean squares for the ANOVA table are the sum of squares divided by the degrees of freedom. Next we have the F statistic for factor A (site) and its p-value. With a p-value of 0.374, the site does not make a difference in tree weight. The row of dots at the upper right indicates the calculator's output has been paused.

```
    DF   SS
A   1   .27225
B   3   7.547
AB  3   .17163
ER  32  10.72668
    F(A)=.81
        P=.374
    F(B)=7.5
```

Press [ENTER] again for the rest of the output. The p-value for tree treatment is 0.001. As we saw before, this does make a difference. Last is the F statistic and its p-value for interaction. There is no interaction in our tree growth example between site and treatment. Lastly, the (pooled) standard deviation is given. Press [ENTER] to return to the first, main menu.

```
B       P=.001
  F(AB)=.17
        P=.915

  S=.5789721496
```

TI-89 Procedure

On a TI-89, for our example, we will enter the data into four lists (one for each tree treatment), and sequentially "block" the sites so that all data for site 1 is before that for site 2.

From the [F6] Tests menu, select option D:ANOVA2-Way, the last item on the menu. You can perform a "blocked" ANOVA, or a 2-Factor with equal observations such as we have. Expand to select the appropriate number of levels of the column factor (4 for the 4 different tree treatments), and type in the number of levels of the row factor (2 for the two different sites). Press [ENTER] to continue.

On this screen enter the lists which were used for the column factors (list1 through list4 in our example). Press [ENTER] to start the calculation and display the first portion of the results.

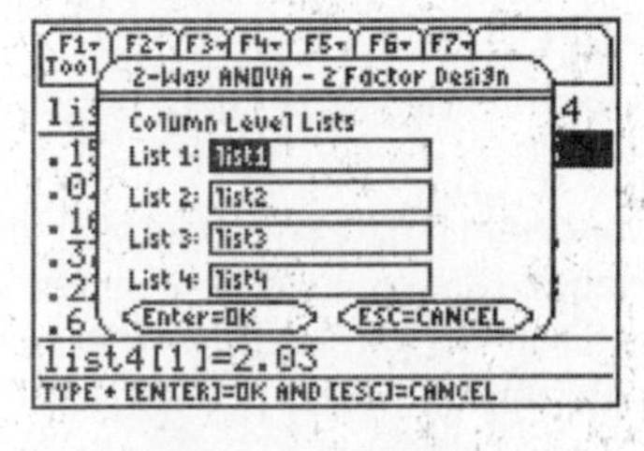

The first portion of the results displays the results for the column factor (our tree treatments). As we saw before, the tree treatment does make a difference in tree weight. Press the down arrow for the rest of the output (row factor, interaction, and error).

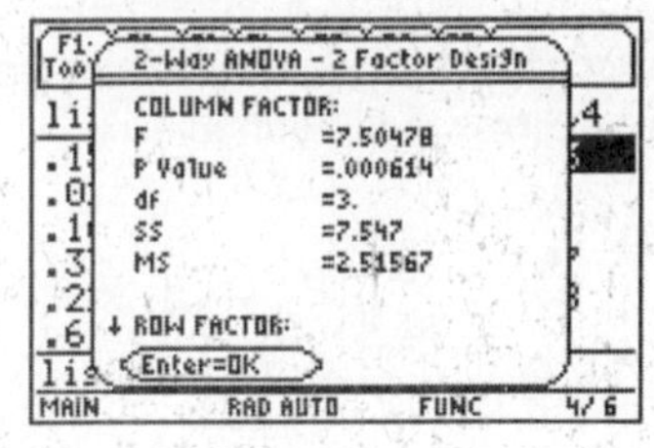

GRAPHING A MEANS PLOT

The text illustrates a graphical way to look for differences in two-way ANOVA with a means plot. The means for the different groups shown in Table 12-5 are displayed here for convenience.

	No Treatment	Fertilizer	Irrigation	Fertilizer and Irrigation
Site 1	0.184	0.332	0.164	1.334
Site 2	0.458	0.630	0.278	1.308

To create a means plot for these data, I have entered the numbers 1 through 4 in L1 to correspond to the four tree treatments, and the means for Site 1 in L2. The means for Site 2 are in L3.

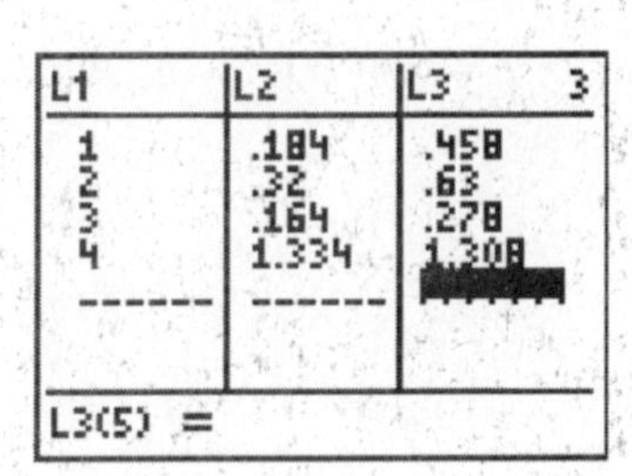

I have defined two connected scatter plots each of which uses `L1` as the `Xlist`. They also use different plot symbols.

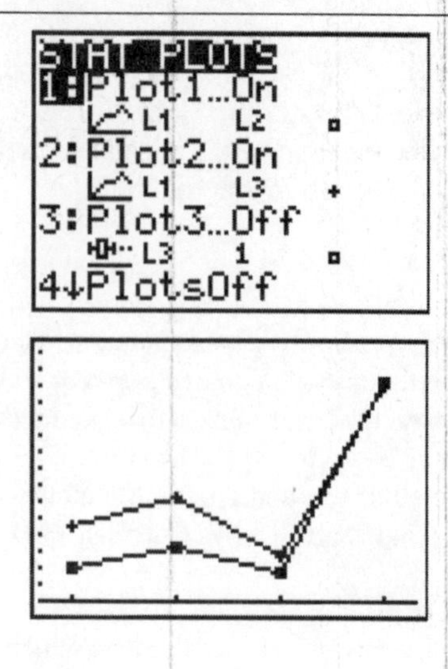

[ZOOM][9] ([F5] on an 89) will display the graph. Significant interactions are shown on these plots through dramatically nonparallel or crossing lines. Significant differences in the second (line) factor are seen by large vertical distances between the two plots. Here, we see the dramatic difference already observed for the combination of fertilizer and irrigation.

Special Case: One Observation per Cell and No Interaction

Our original data on tree growth had five observations for each combination of site and treatment. As an example, assume we have only the first entry in each cell of the table. We thus have only 8 data values (one for each site and fertilizer/water combination). We will not be able to determine an effect due to interaction with only one value per cell, but we can perform two-factor ANOVA and duplicate the text's Minitab results.

Proceed by placing the eight data values into matrix D which will this time be an 8 row, 3 column matrix.

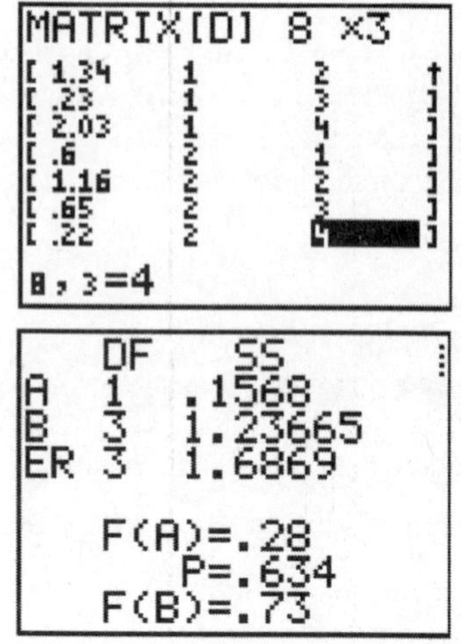

Run program **A1ANOVA** as before. The results shown are similar to the Minitab output displayed in the text.

TI-89 users: proceed analogously using the 2-Way ANOVA as before, except choose `Blocked` as the type of 2-way ANOVA.

13 Nonparametric Statistics

In this chapter we will use the TI calculators' spreadsheet to perform most of the calculations. Examples will be given for each nonparametric test covered in your text. In some circumstances, when sample sizes are too large for the tables in your text, a z-test statistic which is normally distributed can be calculated. The p-value of such a statistic can then be found using the `normalcdf` function. This process will be illustrated.

In all of these tests, the procedures are analogous whether you are using an 83/84 series or an 89 series calculator. If using an 89 series calculator, make the appropriate adjustments to list names.

SIGN TEST

Claims Involving Matched Pairs

EXAMPLE Does the Type of Seed Affect Corn Growth?: The following data was obtained by Louis S. Gossett (the discoverer of the t-distributions) and published in 1908. Two kinds of seed (regular and kiln-dried) were used on *adjacent* plots of land. The listed values are the yields of head corn in pounds per acre. Use a 0.05 significance level to test the claim that there is no difference between the yields from the regular and kiln-dried seed..

Regular	1903	1935	1910	2496	2108	1961	2060	1444	1612	1316	1511
Kiln Dried	2009	1915	2011	2463	2180	1925	2122	1482	1542	1443	1535
Sign of difference	-	+	-	+	-	+	-	-	+	-	-

We let $n = 11$ (disregarding any occurrences 0 differences, of which there were none). If the yields were equal then the number of positive and negative differences would be approximately equal (11/2 = 5.5 each), but in the above table there are only 4 positive differences. Is this significantly different from what we expect? If the two trials are equivalent, the distribution of the number of positive (or negative) signs for the differences would be binomial with $n = 11$ and $p = 0.5$ yielding a mean of $np = 11*0.5 = 5.5$. (You can review the binomial distribution in Ch. 5.)

```
binomcdf(11,.5,4
)
          .2744140625
Ans*2
           .548828125
```

We want the probability of having 4 or fewer positive differences which is P(0) + P(1) + P(2) + P(3) + P(4). Press [2nd] [VARS] for the `DISTR` menu ([F5] on a TI-89) then select `binomcdf(`. Fill in the 11,0.5,4. Press [ENTER] to see the p-value of 0.2744. Since this is a two-tailed test, multiply this p-value by 2 (we get 0.5488). The p-value is much larger than 0.05, so we do not reject the hypothesis that the yields for the two types of seed are equal. The two types of seed were different, but not different enough to be statistically significant.

```
binomcdf(325,.5,
30)
        3.504159E-56
Ans*2
        7.008318E-56
```

Claims Involving Nominal Data

EXAMPLE Gender Selection: The Genetics and IVF Institute conducted a clinical trial of its methods for gender selection. Of 325 babies born to parents using the XSORT method to increase the probability of conceiving a girl, 295 babies were girls. Test the null hypothesis that this method of gender selection has no effect at a 0.05 significance level. Our hypotheses are H_0: p = 0.5 versus H_1: p ≠ 0.5.

This is a binomial distribution with $n = 325$ and $p = 0.5$. The formula in the textbook is really calculating a z-score and then using the normal approximation to the binomial. With TI calculators, we can compute the p-value directly using `binomcdf`. Since there were 295 girls and 30 boys, we want the probability of 30 (or fewer) boys, which we will then double to get the two-sided p-value. The total number of births is $n = 325$. Since the p-value is much smaller than 0.05, there is enough evidence to method does affect the gender of the babies.

Claims about the Median of a Single Population

EXAMPLE Body Temperatures: Use the sign test to test the claim that the median value of the 106 body temperatures of healthy adults (from Data Set 2 of Appendix B) is less than 98.6°F. We are thus testing

H_0: median = 98.6   versus H_1: median < 98.6

The data set has 68 subjects with temperatures below 98.6°F, 23 subjects with temperatures above 98.6°F and 15 subjects with temperatures equal to 98.6°F. (We will later present steps for how to find these values.)

Discounting the 15 temperatures equal to 98.6°F because they do not add any information to this problem, the sample size is $n = 68 + 23 = 91$. If the median were 98.6, we would expect about half of these 91 values to be below the median and half to be above it. This is a binomial distribution with $\mu = np = 91*0.5 = 45.5$. We want the probability of having 23 or fewer values above the median (as this is what has occurred).

Again, we want to select `binomcdf(` from the `DISTR` menu. Fill in the 91,0.5,23 Press ENTER to see the p-value = 0.00000126 (be careful since the leading portion looks like a probability greater than 1). With such a small p-value there is good evidence that there are fewer temperatures above 98.6 than would be expected if it were the median. This supports the claim that the median is less than 98.6.

If the 106 body temperatures are saved in a list (e.g., L1), the following steps show one way the numbers below and above the hypothesized median could be obtained.

With your data in L1, highlight list name L2. Type `L1-98.6` as in the bottom of the screen. Press ENTER to do the calculation. We see some differences are negative, some are positive and some are 0.

Press STAT 2 for the `SortA(` option from the `STAT` menu. Press 2nd 2 to choose L2 for sorting. Then press ENTER. The values will be sorted in ascending order when you go back to the editor.

Use the ▼ key to look at the sorted values. You will see that 68 values are below 98.6 (the differences are negative as at right). You will also find 15 values equal to 98.6 (the differences are 0) and 23 values above 98.6 (the differences are positive.) These are the numbers given earlier.

WILCOXON SIGNED RANK TEST FOR MATCHED PAIRS

EXAMPLE Does the Type of Seed Affect Corn Growth?: The following data are repeated from the first example in this chapter. We will use the Wilcoxon signed-ranks test to test the claim of no difference between type of seed. We use significance level $\alpha = 0.05$.

Regular	1903	1935	1910	2496	2108	1961	2060	1444	1612	1316	1511
Kiln Dried	2009	1915	2011	2463	2180	1925	2122	1482	1542	1443	1535
Difference	-106	20	-101	33	-72	36	-62	-38	70	-127	-24

When we look at the actual differences, we see that some are negative, some positive (there are no zero differences, which would be ignored for testing purposes); but it seems that the negative differences are *larger* than the positive differences. If this is really so, it would seem that the kiln dried seed gives higher yields.

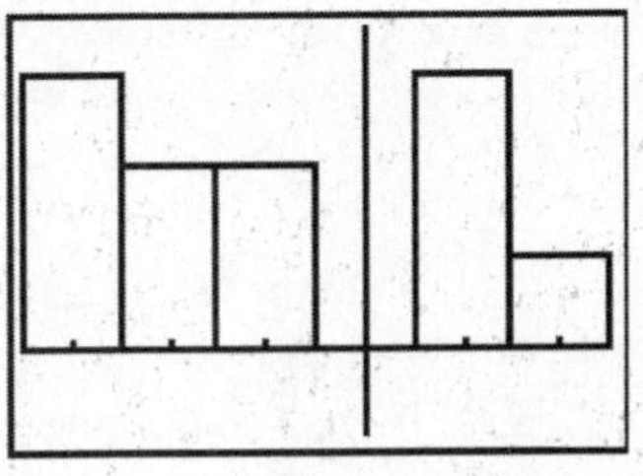

The data are from a matched pairs experiment, since the corn was planted on adjacent fields. From what we learned in Ch. 9, we would like to do a *t*-test using the differences as the data for the test. A histogram of the differences shows that these are not normally distributed. With our small sample sizes, we cannot do the *t*-test. Therefore, we must try a non-parametric test using the differences.

Practically speaking, for this number of data values (11), it might be easier to sort the differences (ignoring signs), assign ranks, and then sum the ranks for the positive and negative differences. The following steps show a method for finding the sum of the positive ranks and the sum of the absolute values of the negative ranks.

```
L2    | L3   | L4    4
2009  | -106 | ------
1915  | 20   |
2011  | -101 |
2463  | 33   |
2180  | -72  |
1925  | 36   |
2122  | -62  |
L4 =abs(L3
```

Put the first sample (regular) in L1 and the second sample (Kiln Dried) in L2. Omit any 0 differences. Highlight list name L3 and type L1-L2 on the bottom line. Press [ENTER] to calculate the differences.

Highlight L4 and enter abs(L3. (The absolute value function is located under the [MATH], NUM(ber) menu. It is option 1 on a TI-83/84, and option 2 on a TI-89.) Press [ENTER]. Now L4 will contain all positive values (the absolute value of the differences).

```
L3    | L4   | L5    5
-106  | 106  | -106
20    | 20   | 20
-101  | 101  | -101
33    | 33   | 33
-72   | 72   | -72
36    | 36   | 36
-62   | 62   | -62
L5(1)=-106
```

Copy the differences in L3 into L5 by highlighting L5, entering L3, and then press [ENTER].

```
L3    | L4   | L5    5
-106  | 20   | 20
20    | 24   | -24
-101  | 33   | 33
33    | 36   | 36
-72   | 38   | -38
36    | 62   | -62
-62   | 70   | 70
L5(1)=20
```

Press [STAT] [2] to paste SortA(on the home screen. Then enter L4,L5 and press [ENTER] to see Done. You have sorted L4 and carried the contents of L5 along. Press [STAT] [1] to return to the STAT editor. L5 retains the sign of the original differences. We want to assign ranks to the differences based on the absolute values in L4, but retain the signs which are carried (for now) in L5.

Highlight L5 and enter L5 [÷] L4. Press [ENTER] for a column of positive and negative ones as at right.

Generate the integers from 1 to 14 in L6. Do so by highlighting L6 and typing [2nd] [STAT] [▸] [5]. This chooses the seq(option from the LIST, Ops menu. (On a TI-89, press [F3] for the Lists menu, then [2] for Ops, then [5]). Type X,X,1,14). Press [ENTER]. At this point, you will need to scan down L4 and search for any ties. This particular data set has none, but if there are any, you would need to modify L6 (the ranks) to assign an average rank for any differences which are equal. For example, if the second and third values in L4 were both 3, they would each receive a rank of (2+3)/2 = 2.5.

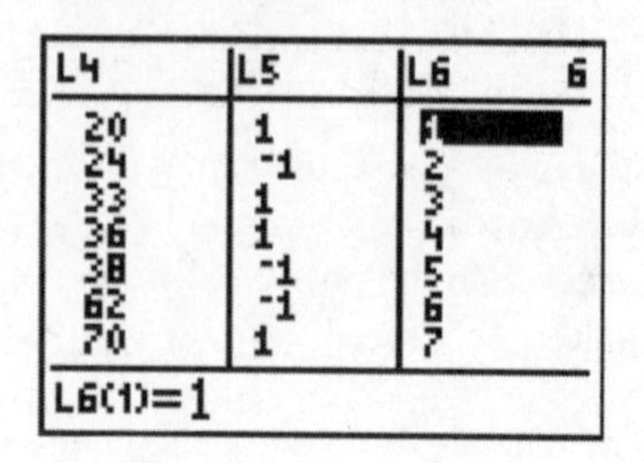

Quit and return to the Home screen. Type 11(11+1)/2 [STO▸] [ALPHA] [MATH] [ENTER] for the sum of the ranks (66) to be stored as A (Variable a on a TI-89 is [alpha][=]). As a double-check on your (possibly adjusted) ranks in L6, sum the elements in L6 with [2nd] [STAT] [▸] [▸] [5] [2nd] [6] [ENTER] (on a TI-89 from the home screen, press [2nd][5] for the MATH menu, arrow to List, press the right arrow to expand the selection and choose option 6. This too should be 66. If not, then recheck your ranks in L6.

Multiply L5*L6 and store in L6. This puts a sign on the rankings in L6.

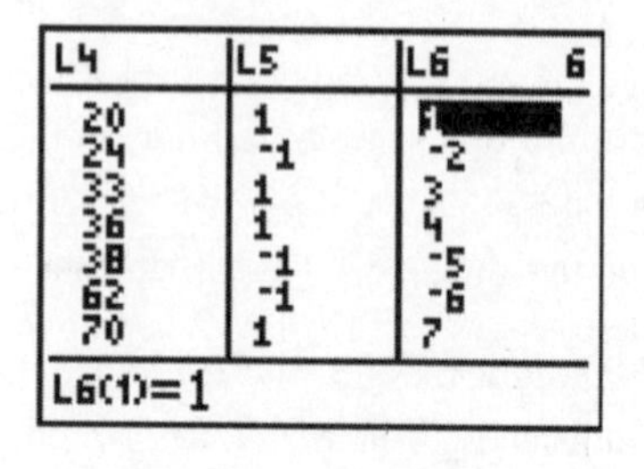

You can (carefully) add the negative and positive ranks manually, or use this procedure. Remember, we stored the total sum of ranks in A. Adding the signed ranks in L6 to the sum of all ranks will leave us with only the positive ranks (times two). Any negative ranks will "cancel out" their positive counterparts. So (on the home screen) if we enter (sum(L6)+A)/2 [ENTER], we find the sum of the positive ranks is 15. Since we know the sum of all the ranks is 66 (from above), we find that the sum of the negative ranks is 51.

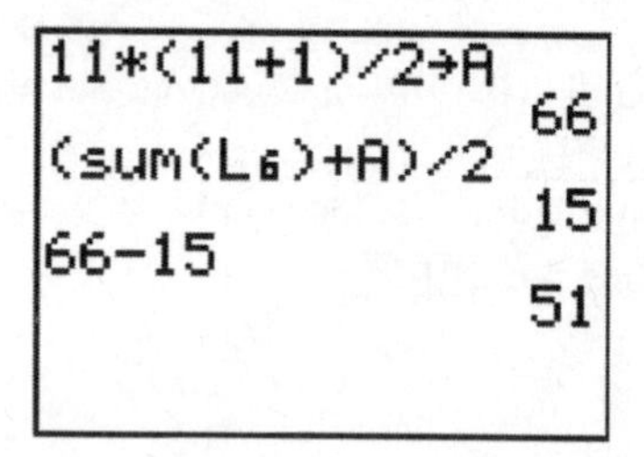

We use the smaller value (15) for comparison with the appropriate value from the table in the text. (Table A-8). We find it is larger than the critical value (11), we cannot reject H_0. This test has failed to detect a difference in yield between regular and kiln dried corn.

WILCOXON RANK-SUM TEST FOR TWO INDEPENDENT SAMPLES

EXAMPLE BMI of Men and Women Data Set 1 in Appendix B of your text includes Body Mass Index (BMI) scores for both men and women. We will use the first 13 sample values for men and the first 12 for women. The data are reproduced below for convenience. (Only a part of the complete data set is used so that the calculations are easier to follow). Use these two sets of independent sample data with a 0.05 significance level to test the claim that median BMI scores for men and women are the same.

Men	23.8	23.2	24.6	26.2	23.5	24.5	21.5	31.4	26.4	22.7	27.8	28.1	25.2
Women	19.6	23.8	19.6	29.1	25.2	21.4	22.0	27.5	33.5	21.6	29.9	17.7	

The following steps give the sums of ranks for the two data sets.

Put all 25 of the BMI scores into L1. In L2 type a 1 next a man's score and a 0 next to a woman's score. You will thus have thirteen 1's followed by twelve 0's in L2.

Make a copy of L1 in L3 and a copy of L2 in L5. (Highlight the name of the receiving list and enter the original list's name, then press [ENTER].)

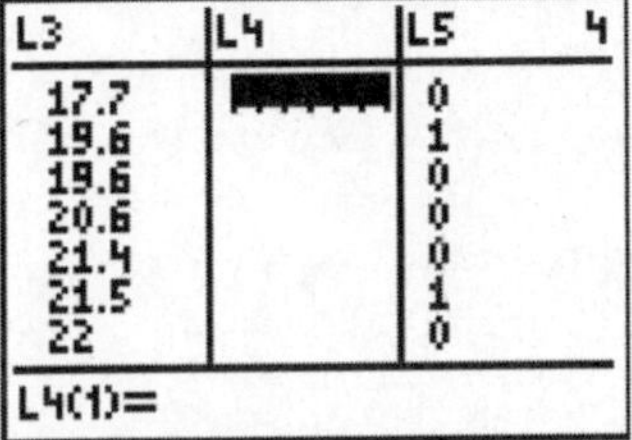

Press [STAT] [2] to paste SortA(on the home screen, then type L3,L5 [ENTER] to sort the values in L3 and carry along the values in L5. On a TI-89, use this command on the home screen and locate the command on the Math, List menu ([2nd][5][3][4]), then use [VAR-LINK] to enter the list names. The body mass values have been sorted from lowest to highest, and the gender indicator has been preserved.

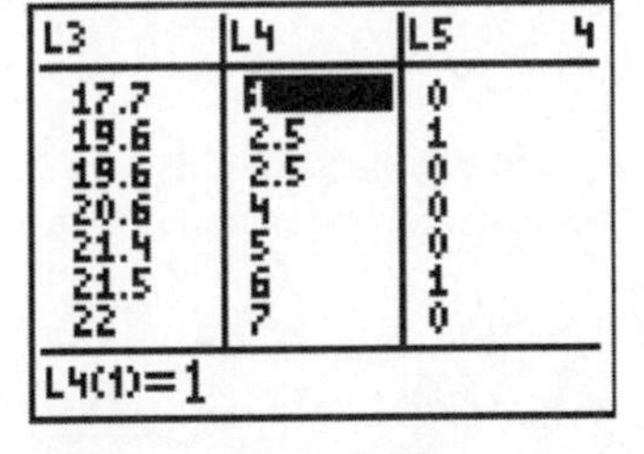

Generate the integers from 1 to 25 in L4 as follows. Type [2nd] [STAT] [▶] [5] to choose the seq(option from the LIST, Ops menu. Type X,X,1,25 and [ENTER]. Next, modify L4 as necessary, so that it has the ranks of the values in L3. Make sure to handle tied values by giving each a rank of the average of the ranks of the tied values. In the screen at right, there are two values of 19.6, so each is assigned a rank of 2.5. There are three data values with ties in our data set.

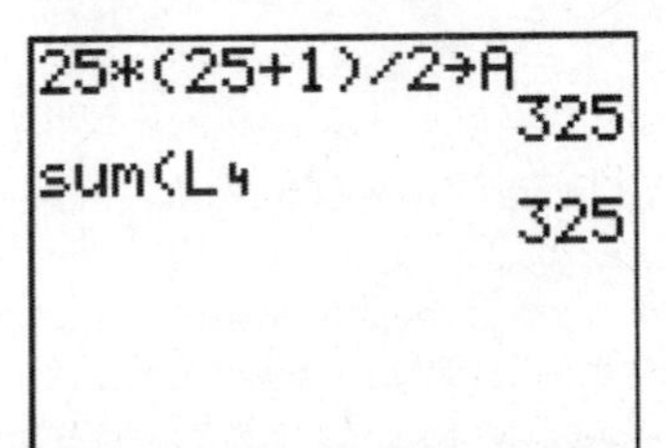

Quit and return to the Home screen. Type 25(25+1)/2 [STO▶] [ALPHA] [MATH] [ENTER] for the sum of the 25 ranks which is 325, which is also stored into variable A. Find sum(L4) which should also be 325.

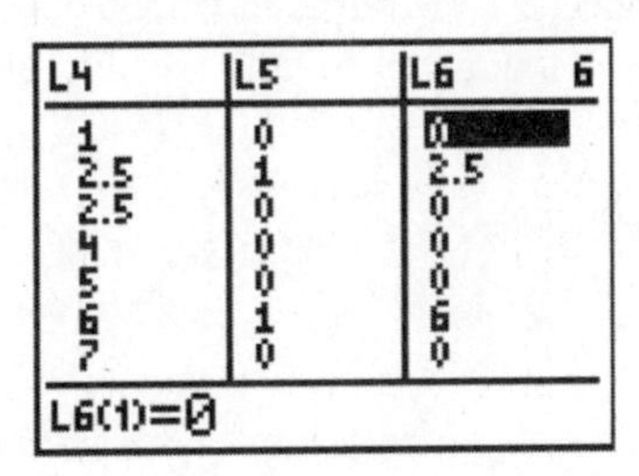

Multiply L4 by L5 and store in L6. L6 now contains the BMI rankings for the men. (Women have a 0 in L5.)

Find `sum(L6` as at right. This is the sum of the men's BMI ranks or $187 = R_1$ in the text. The sum of the women's BMI ranks is found by subtracting this value from the sum of all ranks. This value is 138.

```
sum(L6
                187
325-187
                138
```

The above value for R is compared against a value from a normal distribution with mean $\mu = \frac{n_1(n_1+n_2+1)}{2} = 169$ and standard deviation $\sigma = \sqrt{\frac{n_1 n_2(n_1+n_2+1)}{12}} = \sqrt{\frac{13*12(13+12+1)}{12}} = 18.835$. The Z-score of this statistic is found by the formula $Z = \frac{R-\mu_R}{\sigma_R} = \frac{187-169}{18.385} = 0.98$. We find the area in the right tail of the standard normal distribution past $Z = 0.98$ by pressing [2nd] [VARS] [2] to paste `normalcdf(` on to the home screen and then typing .98,E99 and pressing [ENTER]. We see the result is .1635. We double this for the two tailed test p-value of 0.3271. We do not have evidence to reject the hypothesis that the distributions for the two samples were the same. It appears that men and women have similar body mass indexes.

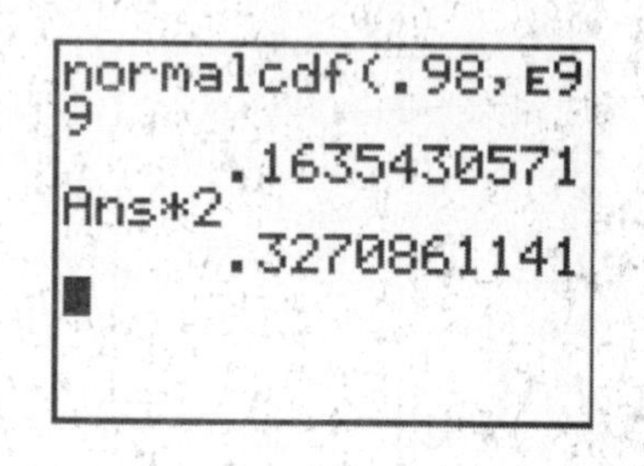

KRUSKAL-WALLIS TEST

The Kruskal-Wallis test is an extension of the Wilcoxon Rank-Sum test. It compares the distributions from more than two independent samples.

EXAMPLE Effects of Treatments of Poplar Tree Weights: Table 13-6 (reproduced below for convenience) lists weights of poplar trees given treatments. Use the Kruskal-Wallis Test to test the null hypothesis that the samples all came from populations with the same distribution. Use significance level 0.05.

None	.15	.02	.16	.37	.22
Fertilizer	1.34	.14	.02	.08	.08
Irrigation	.23	.04	.34	.16	.05
Fertilizer and Irrigation	2.03	.27	.92	1.07	2.38

The steps below will explain how to get the sum of the ranks for the data from a specific sample.

Place all of the data values in `L1`. Start with the values for trees with no treatment, followed those who got fertilizer, etc. In `L2`, place a 1 next to each of the values that got no treatment, a 2 next to each Fertilizer value, a 3 next to each Irrigation value, and a 4 next to the values for trees that received both fertilizer and irrigation. Make a copy of `L1` in `L3`.

L1	L2	L3
.15	1	.15
.02	1	.02
.16	1	.16
.37	1	.37
.22	1	.22
1.34	2	1.34
.14	2	.14

L3(1)=.15

Press [STAT] [2] to paste `SortA(` onto the home screen, then type `L3,L2` [ENTER] to sort the values in `L3` carrying along `L2`.

Generate the integers from 1 to 20 in `L4` as follows. Type [2nd] [STAT] [▶] [5] to choose the `seq(` option from the `LIST`, `Ops` menu. Type X,X,1,20 and [ENTER]. Next, modify `L4`, so that it has the ranks of the

L2	L3	L4
1	.02	1.5
2	.02	1.5
3	.04	3
3	.05	4
2	.08	5.5
2	.08	5.5
2	.14	7

L4(1)=1.5

values in L3. Make sure to handle ties by assigning the average of the ranks to all tied observations. Here, we show that the sum of ranks 1 to 20 is 210 and also confirms that the sum of the ranks in L4 is also 210. This is a double-check that the ranks were assigned properly.

```
20*(20+1)/2
                210
sum(L4
                210
```

Next perform SortA(L2,L3,L4 to sort the values in L2 and carry along the values in L3 and L4.

In L5, place five 1's followed by fifteen 0's. This places a 1 next to each non-treatment data value.

Multiply L4 by L5 and store the results in L6. This stores only the non-treatment ranks in L6. Then find sum(L6) (Find sum(on the LIST,MATH menu) to obtain $R_1 = 45$.

```
L4      L5      L6     6
9.5     1       9.5
15      1       15
11      1       11
8       1       8
1.5     1       1.5
7       0       0
5.5     0       0
L6(1)=9.5
```

Repeat steps as above but with 1 next to the irrigation ranks and 0's next to all others to get R_2 and similarly with for the other two treatments. Once you have determined that $R_1 = 45$, $R_2 = 37.5$, $R_3 = 42.5$ and $R4 = 85$, you can calculate the test statistic H by the formula $H = \frac{12}{N(N+1)}\left(\frac{R_1^2}{n_1}+\frac{R_2^2}{n_2}+\frac{R_3^2}{n_3}+\frac{R_4^2}{n_4}\right)-3(N+1)$

$$= \frac{12}{20(20+1)}\left(\frac{45^2}{5}+\frac{37.5^2}{5}+\frac{42.5^2}{5}+\frac{85^2}{5}\right)-3(20+1) = 8.214.$$

```
)
        71.21428571
Ans-3*21
        8.214285714
χ²cdf(8.214,E99,
3)
        .0417899399
```

The distribution of H is chi-squared with $k - 1 = 4 - 1 = 3$ degrees of freedom. We find the p-value of 8.214 using the χ^2cdf(function from the Distributions (Distr) menu. On the home screen, press 2nd VARS for the Distr menu, and select χ^2cdf(to paste it on the screen. Then type 8.214, E99,3 and ENTER. (Find the Distr menu by pressing F5 on the Statistics/List Editor on a TI-89). We find the p-value is 0.0417. This is less than the significance level of 0.05, so we have evidence against the null hypothesis that the samples came from populations with the same distribution. At least one of the treatments has made a difference in weight of the trees.

RANK CORRELATION

EXAMPLE Student and *U.S. News and World Reports* Rankings of Colleges: Table 13-7 (reproduced below for convenience) includes rankings for several colleges by students and *U.S. World and News Reports* magazine. Test to see if there is a correlation between the rankings given by students and the magazine. Use significance level $\alpha = 0.05$.

College	*Student Ranking*	*Magazine Ranking*
Harvard	1	1
Yale	2	2
Cal. Inst. Of Tech.	3	5
M. I. T.	4	4
Brown	5	7
Columbia	6	6
U. Penn.	7	3
Notre Dame	8	8

With the two sets of ranks in L1 and L2, the scatter diagram shows evidence of a linear relationship between the ranks given by students and the magazine (with an outlier). From the STAT TESTS menu, choose option LinRegTTest. Use the two lists as the data for the linear regression. Highlight Calculate in the last line, and press [ENTER].

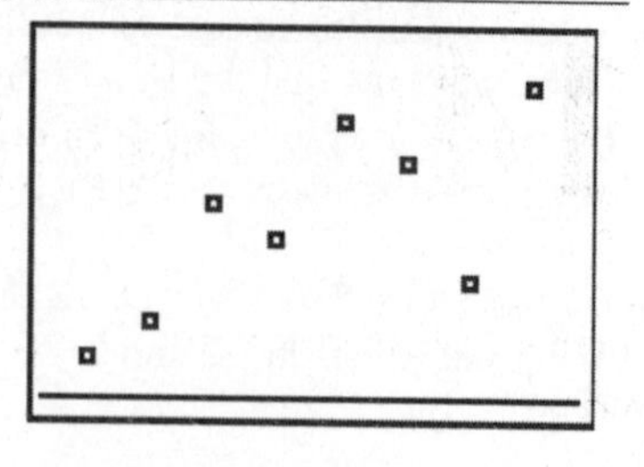

The last line of output from the regression gives $r_S = 0.714$.
To test the hypothesis, use the critical value from Table A-9 of the text. Do not use Table A-6 (which was used for the Pearson correlation coefficient) because it requires the populations sampled to be normally distributed.

```
LinRegTTest
 y=a+bx
 β≠0 and ρ≠0
↑b=.7142857143
 s=1.8516402
 r²=.5102040816
 r=.7142857143
```

The critical value is $0.738 > 0.714$, so we fail to reject the null hypothesis. It appears that students and the magazine have different perceptions of colleges.

Note: The *t* statistic and p-value given in the regression output are for the Pearson correlation coefficient and do not apply for the ranks.

RUNS TEST FOR RANDOMNESS

This is a test to decide whether (or not) data actually were randomly obtained. It is based on whether there are too few (or too many) runs of like observations in a dichotomous (only two outcomes possible) data set.

EXAMPLE Boston Rainfall on Mondays: Refer to the rainfall amounts for 52 consecutive Mondays in Boston as listed in Data Set 11 of Appendix B and repeated below with 0 representing no rain and 1 representing some rain. Is there sufficient evidence to support the claim that rain on Mondays is not random. Use a 0.05 significance level.

0000 1 0 1 00 1 00 1 000 1 00 111 0000
1 0 1 0 111 0 1 000 1 000 1 0 1 00 1 000 1

$n_1 = 33$ = the number of zeros or dry days.
$n_2 = 19$ = the number of ones or days with some rain
$G = 30$ = the number of runs or groupings above

Since $n_1 > 20$ we can use formulas given in the text for

$$\mu_G = \frac{2n_1n_2}{(n_1+n_2)} + 1 = \frac{2*33*19}{(33+19)} + 1 = 25.115 \text{ and } \sigma_G =$$

$$\sqrt{\frac{(2n_1n_2)(2n_1n_2 - n_1 - n_2)}{(n_1+n_2)^2(n_1+n_2-1)}} = \sqrt{\frac{(2*33*19)(2*33*19-33-19)}{(33+19)^2(33+19-1)}} = 3.306.$$

```
(2*33*19/(33+19)
+1
        25.11538462
√((2*33*19)*(2*3
3*19-33-19)/(52²
*51)
        3.306074002
```

Both formulas use only the values for n_1 and n_2. The test statistic z is computed as $z = (G-\mu_G)/\sigma_G = (30-25.115)/3.306 = 1.48$.

We use normalcdf(pasted in from the DISTR menu to compute the p-value for the test. We double the p-value for the two-tailed test and obtain $0.1389 > 0.05$. Thus we fail to reject the null hypothesis.
There is no evidence that rainfall on Mondays in Boston is not random.

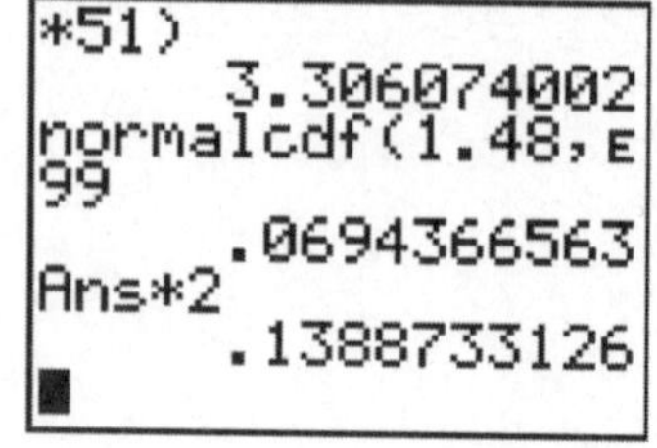

14 Statistical Process Control

In this chapter we will use TI calculators to plot run charts and control charts for the range, $\bar{x}$, and proportions.

As with nonparametric tests, the procedures using a TI-89 are analogous. If you have forgotten how to do these connected scatter plots, refer to Chapter 2.

RUN CHARTS

EXAMPLE Manufacturing Aircraft Altimeters: Treating the 80 altimeter errors in Table 14-1 of the text as a string of consecutive measurements, construct a run chart using a vertical axis for the errors and a horizontal axis to identify the order of the sample data.

Put the 80 data values in list L1. Using 1-Var Stats from the STAT, CALC menu, we find the overall mean is 6.45. Now, generate integers from 1 to 80 and store them in L2. Highlight the list name and use the seq(option from the LIST, OPS menu and complete the command by typing X,X,1,80).

Set up Plot1 as an xy-Line plot. All other statplots should be turned off.

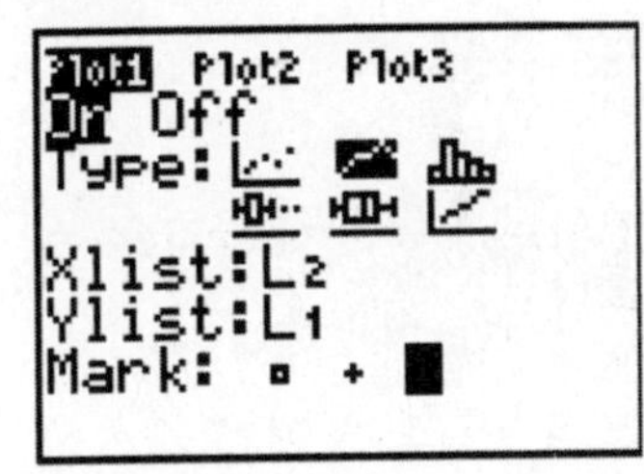

Press [Y=] and let Y1 = 6.45.

Temporarily turn off the axes by pressing [2nd] [ZOOM] and choosing AxesOff. (Be sure to change this back after you are through with this chapter).

Press [ZOOM] [9] then adjust the Window to better fill the screen by setting Xmin = 0 and Xmax = 80. Then press [GRAPH] to display the resized plot. To maneuver within the plot, press [TRACE] and use the right and left arrows. The reference line for the mean of the observations is clear, as is the increasing variability as time continues.

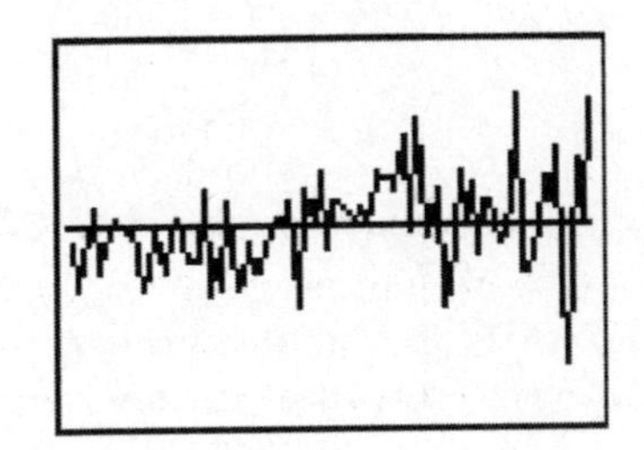

Below is part of Table 14-1 which will be useful in the next two examples.

Day	Mean	Range	Day	Mean	Range	Day	Mean	Range	Day	Mean	Rnnge
1	2.50	19	6	-0.75	26	11	10.75	3	16	9.50	17
2	2.75	13	7	0.00	21	12	12.75	11	17	8.75	10
3	1.00	15	8	0.25	13	13	21.00	13	18	8.25	43
4	0.25	11	9	1.25	26	14	13.00	28	19	12.25	31
5	2.00	11	10	10.50	18	15	3.25	32	20	9.75	63

CONTROL CHART FOR MONITORING VARIATION: THE R CHART

EXAMPLE Manufacturing Aircraft Altimeters: Refer to the altimeter errors in Table 14-1 of the text. Using the samples of size $n = 4$, construct a control chart for R

Put the day (integers 1 to 20) in L1 and the corresponding ranges in L2. To find the mean of the data in L2 use 1-Var Stats from the STAT, CALC menu. You will find $\overline{R} = 21.2$. You can use this value to find the control limits as in the text using the multipliers in Table 14-2. You will find UCL = 48.4 and LCL = 0.

Set up Plot1 to be an xy-line plot as at right.

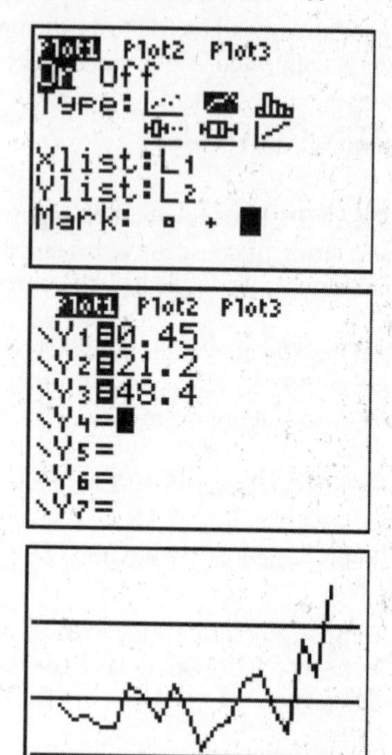

Set up the Y= Editor with the control limits as calculated.

Press ZOOM 9 then press TRACE for the plot which is similar to the R-chart in the text. This plot confirms the indication from the runs chart – the process is getting out of control.

CONTROL CHART FOR MONITORING PROCESS MEAN: THE $\overline{x}$ CHART

EXAMPLE Manufacturing Aircraft Altimeters: Refer to the altimeter errors in Table 14-1 of the text. Using the samples of size n=4, construct a control chart for $\overline{x}$.

Put the day (integers 1 to 20) in L1 and the corresponding means in L2. To find the mean of the data in L2, use 1-Var Stats from the STAT, CALC menu. You will find $\overline{\overline{x}} = 6.45$. You can use this value and that of $\overline{R}$ (found above) to find the control limits as in the text, using the constants from Table 14-2. You will find UCL = 21.9 and LCL = –9.0.

Set up Plot1 to be an xy-line plot, as has been done previously.

Set up the Y= Editor with the control limits as was done for the range plot above.

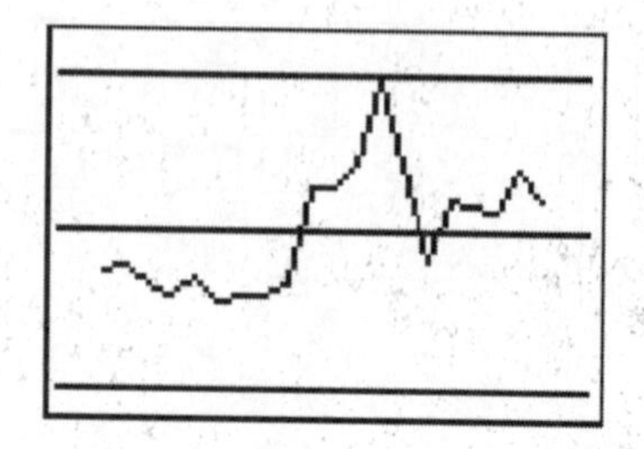

Press ZOOM 9. Change the Window so that the LCL is visible. Try using Ymin = –10. Then press GRAPH to display the plot which is similar to the $\overline{x}$-chart in the text.

CONTROL CHART FOR ATTRIBUTES: THE P-CHART

EXAMPLE Defective Aircraft Altimeters: In this chapter, we have been studying aircraft altimeters. An altimeter is considered defective if it cannot be calibrated or corrected to give accurate readings. Listed below are the numbers of defective altimeters in successive batches of 100. Construct a control chart for the proportion p of defective altimeters and determine whether the process is within statistical control.

Defects	2	0	1	3	1	2	2	4	3	5	3	7

Put the batch number in L1 (integers from 1 to 12) and the corresponding number of defectives in L2.

On the Home screen, type `L2/100` and store in L2. This will replace the numbers of defectives with the proportion of defectives.

Set up `Plot1` as in the preceding examples.

Set up the Y= editor with the control limits as has been done previously. These limits are calculated as $\overline{p}$ (use 1-Var Stats to find the average of the proportions, which is 0.0275), and $\overline{p} \pm 3\sqrt{\frac{\overline{pq}}{n}}$ which give 0.0766 and -0.0216 (use 0 for any values less than 0).

Press [ZOOM] [9]. Change `Ymin` and `Ymax` if necessary so that both the LCL and UCL are visible. Use [TRACE] as necessary to fully examine the graph.

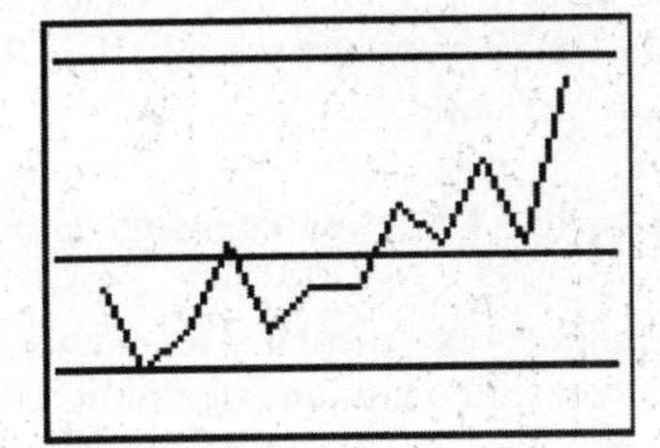

Appendix

LOADING DATA APPS OR PROGRAMS FROM A COMPUTER OR ANOTHER TI-83/84

This appendix contains information on transferring data and programs from the CD-ROM that came with your main text and is available from the publisher. Also, we give instruction on how to install a data set from the data Apps and how to make and install your own group.

The CD-ROM contains the data sets from Appendix B of *Elementary Statistics* (10th Edition) by Mario F. Triola. For the TI-83+ and TI-84 calculators, data is given in an App (or application) called TRIOLAXE.APP. For the TI-83/89, data is given as individual lists in ASCII format as text files (with extensions of .txt). The CD-ROM also contains two programs A1ANOVA.83p (used in Chapter12) and A2MULREG.83p (used in Chapter 10).

Your instructor will probably load the data (and programs, if needed) onto your TI calculator, or you can transfer them from the CD-ROM with your computer if you have TI-Connect or TI-GRAPH LINK software and cable available from Texas Instruments (the cable and software are included with the TI-84, as well as the TI-89 Titanium edition). See the guidebook that comes with the calculator for information on the TI-GRAPH LINK.

Loading Data or Programs from One TI-83/84 to Another

Connect one calculator to another with the cable that came with the calculator. The link port is located at the center of the bottom edge of the calculator on the 83 series, on the top for the 84 series (which can communicate with 83's with the serial cable, and with other 84s with the USB cable as well).

On the receiving calculator, press [2nd] [X,T,Θ,n] to choose the LINK menu. Press [▶] to highlight RECEIVE. Press [ENTER] or [1] and see the message Waiting displayed.

SEND RECEIVE
1:All+...
2:All-...
3:Prgm...
4:List...
5:Lists to TI82...
6:GDB...
7↓Pic...

On the sending calculator, press [2nd] [X,T,Θ,n]. Press the down arrow to scroll through the list. When the category if what you wish to send is highlighted, press [ENTER]. In this example we are sending choice C: Apps, but we could send a program or a list of data with options 3 and 4. Once we press [ENTER] to select option C: Apps, we use [▼] to locate the TRIOLAXE App and press [ENTER] to select it.

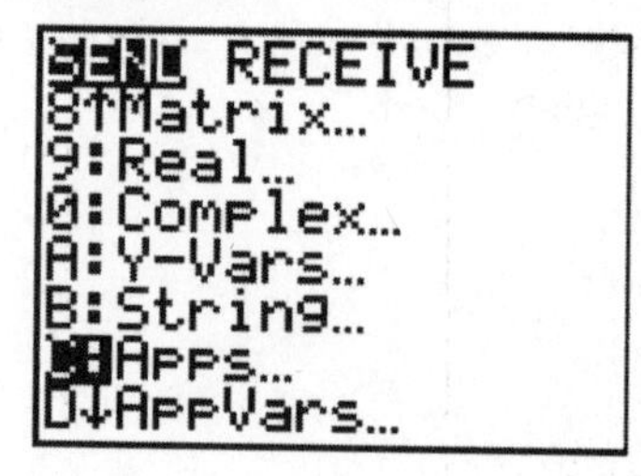

Press [▶] to highlight TRANSMIT. Press [ENTER] or [1] to transmit whatever you have chosen on the previous step.

Note: If you are transmitting an App the receiving calculator will signal "garbage collecting" then "receiving", and then "validating" before indicating "Done", so be patient. If the name of whatever is being sent is already in use on the receiving calculator that calculator will show a screen like the one at right. You can then choose to overwrite the old with the new or to rename so you can keep both.

```
DuplicateName
1:Rename
2:Overwrite
3:Overwrite All
4:Omit
5:Quit
 FPULS     LIST
```

USING THE DATA APPS (TI-83+/84 ONLY)

Press the [APPS] key for a list of all applications which have been installed on the calculator. (We assume that TRIOLAXE App has been installed in this example. The list of other Apps available may vary.) Use [▼] to locate the TRIOLAXE App.

Then press [ENTER] for a title screen that soon changes to this screen which contains a list of the data sets in Appendix B of your text.

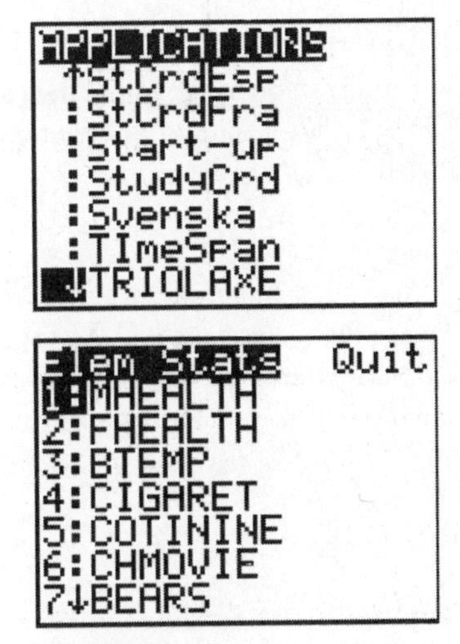

Press [2] to choose FHEALTH and see a list of the data lists from the data set Fhealth.

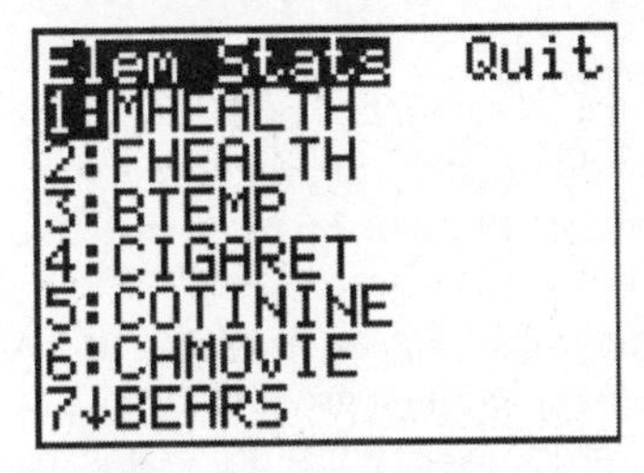

Cursor down and press [ENTER] next to each list you wish to load. A small black square will designate a chosen list. For our example we choose FAGE, FWT and FPULS. If you wish to load all lists in a set then press [▶] to highlight "All".

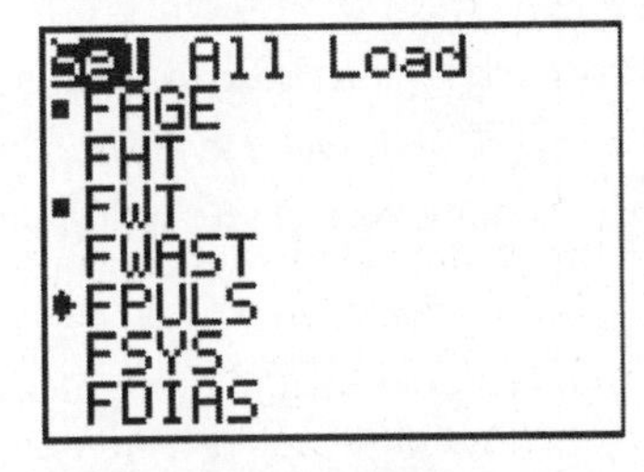

Press [▶] [▶] to highlight Load. At this screen, press [1] to choose SetUpEditor.
Note: The option 2:Load loads the lists from archive memory to random access memory. The data would not be loaded to your Stats Editor but would be available on the Lists menu by pressing [2nd] [STAT].

At this screen, press [2] to choose the "Exchange Lists" option.
Note: Option 1 will add the lists to the lists already in the Stats Editor. Option 3 will not change the editor and place the lists in the LIST menu as described in the Note above.

```
Sel All Setup
1:Add to Editor
2:Exchange Lists
3:No Change
```

After making your selection, you will be returned to the previous screen. Now, select option 2:Load.

This screen informs you that the List(s) have been loaded. Pressing any key returns you to the initial data set selection screen. Pressing [2nd] [MODE] lets you QUIT and return to the Home screen (or press the right arrow and then [ENTER]).

```
List(s) have
been loaded.
            Done

PRESS ANY KEY
```

To see the results, press [STAT] [1] to check out the Stat Editor. Here we see the lists we loaded are in fact there and ready for us to use.

FAGE	FWT	FPULS 1
17	114.8	76
32	149.3	72
25	107.8	88
55	160.1	60
27	127.1	72
29	123.1	68
25	111.7	80

FAGE(1) =17

GROUPING AND UNGROUPING (TI-83+/84 ONLY)

Just as the Data App has lists of data from the text grouped together, you may want to group lists of data and/or matrices and/or programs. The advantage to doing this is that groups are saved in Archive memory and do not take up room in active RAM memory until you want to use them.

Example Group the data sets FAGE, FWT, and FPULS just moved in from the App in the preceding example into a group called FEM. (These data are already part of the group FHEALTH in the TRIOLAXE App, so we are just doing this to provide an example.)

Press [2nd] [+] to choose the MEM menu. Then choose option 8:Group. Press [ENTER] to select to create a new group. You will be prompted to name your group. Type in the name FEM and press [ENTER].

```
GROUP UNGROUP
1:Create New
```

You are now asked what should be in the group. Press [4] to choose List.

```
GROUP
1:All+...
2:All-...
3:Prgm...
4:List...
5:GDB...
6:Pic...
7↓Matrix...
```

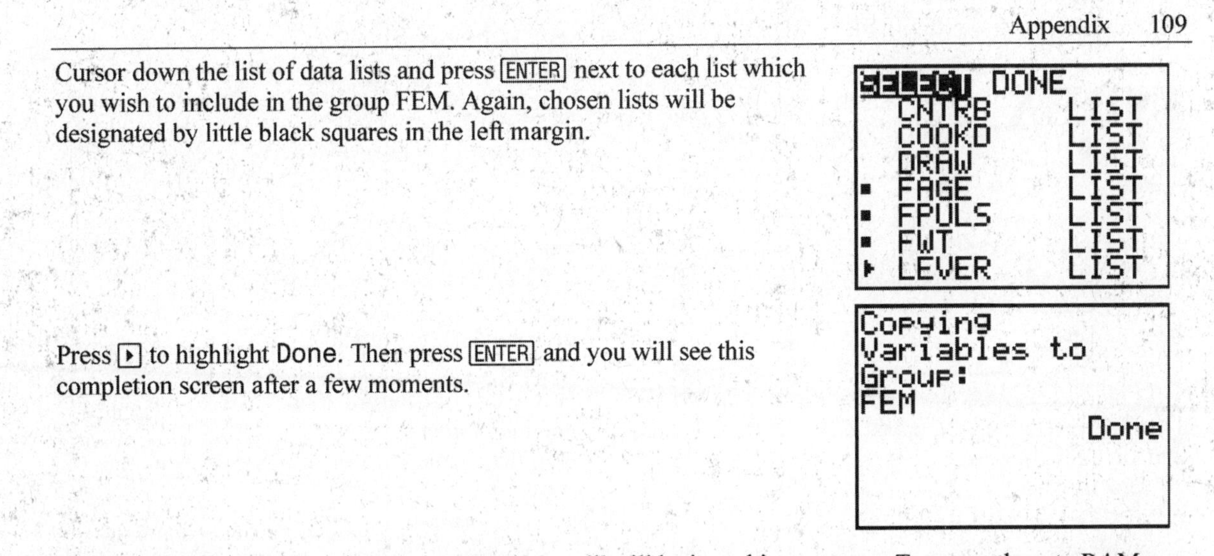

Cursor down the list of data lists and press [ENTER] next to each list which you wish to include in the group FEM. Again, chosen lists will be designated by little black squares in the left margin.

Press [▶] to highlight `Done`. Then press [ENTER] and you will see this completion screen after a few moments.

Now if you delete the three lists from RAM they will still be in archive memory. To return them to RAM, return to Group/Ungroup screen (option 8 from the `MEM` menu) and highlight `Ungroup` and press [ENTER]. You would then select group FEM to ungroup. Even though the list will then reside in RAM, it will remain in archive memory unless it is deleted as a Group using option `2:Mem Mgmt/Del` from the `MEM` menu.

USING .TXT FILES (ALL MODELS)

The CD-ROM which is included with the text has data files in text (.txt) format which can be loaded into the calculators without retyping the data, if you have the proper computer cable and TI-Connect software. While using this procedure is probably not time effective for small data sets, it can save both time (and the aggravation of typing mistakes) with larger ones. The method is simple for anyone who is familiar with basic cut-and-paste editing on a computer.

Insert the CD, and select the Datasets folder. Then, select the textfiles folder. There are icons and short names for each data set (list). Select the data set you wish and double click it to open it (the computer will use Notepad as the default software). Drag the mouse cursor to highlight the list contents. Press Control-C to copy the data onto the computer's scratchpad.

Start the TI-Connect software, and select the TI DataEditor application. Click on the blank page icon. You will see a list with a zero (Ø) as the first entry. Click the cursor in the first cell, and press Control-V (Paste) to place the list contents into the editor. You will need to give your list a "name" and TI-calculator properties. Click File, Properties (Control-R). Select the appropriate device type and give the list a name. If you are using a TI-89 series calculator, you will also be asked the folder you want the data placed in. Click the OK button. Now, click the Send File icon on the menu bar (it looks like a TI calculator with an arrow pointing toward it); the transfer should start automatically. If the list name is already in use (`L1` through `L6`, for example), you will be asked whether you wish to abort the transfer, overwrite the current contents, or give a new name to the list.

INDEX